My Beaver Colony

My Beaver Colony

BY LARS WILSSON

Translated from the Swedish by Joan Bulman

SOUVENIR PRESS
London

First published in Sweden by
Albert Bonniers Forlag under the title of *Baver*

First published in Great Britain 1969
by Souvenir Press Ltd., 95 Mortimer Street,
London, W.1 and simultaneously in Canada
by The Ryerson Press, Toronto 2, Canada

SBN: 285 50221 2

*All photographs except Figs. 7 and 28
are by the author*

Made and printed by offset in Great Britain
by William Clowes and Sons, Limited,
London and Beccles

Preface

The beaver belongs to the order of rodents, a large group of animals that includes almost half of all species of mammals. All rodents have a number of characteristics in common—not merely physical characteristics but also movements and behavior patterns. Particularly characteristic are the two pairs of long, chisel-shaped front teeth, which continue to grow all through the animal's life and have enamel only on the front. In the beaver, as in many other rodents, this enamel is an attractive orange-red.

The family to which the beaver belongs (*Castoridae*) occupies in many respects a unique position among rodents. It now comprises only two species, the European beaver (*Castor fiber L.*) and the American (*Castor canadensis Kuhl.*), which are so closely related that they probably ought to be regarded as a single species.

The adult beaver may weigh as much as fifty-five to sixty-five pounds and among rodents is exceeded in size only by the South American species. Characteristic of the beaver are the peculiar tail and two protuberances the size of a clenched fist at the opening of the genital canals, known as the castoreum pouches, which, in common with the genital, intestinal and urinary canals open into a pocket of skin that is generally, rather incorrectly, referred to as the "cloaca."

But it is above all the beaver's behavior that is most peculiar among rodents. The beaver is highly specialized for life in and beside water and he is a skillful swimmer with the ability to remain under water up to fifteen minutes at a time.

During the summer he lives exclusively on plants, leaves and a certain amount of bark from deciduous trees. In colder parts of the world he spends the winter under the covering of ice and inside his underground holes and passages, all of which have their openings under the water. He is reduced to living entirely on the bark of the branches he has laid by during the autumn as a winter store. While working on his winter store the animal fells large deciduous trees. When the branches have been stripped of bark, they are used as building material.

The beaver's ability to build lodges, which may be up to ten feet high, and to regulate the water level by means of dams that are often several feet high and several hundred yards long and give rise to large lakes, is unique in the animal world. Yet perhaps the most interesting thing of all is the beaver's very highly developed family life.

Contents

Preface v

Introduction ix

A Little Beaver History 1

The Beaver and Primitive Man 2

The Beaver and Civilization 3

The Great Beaver Slaughter 4

The Beaver Returns 6

Elias and the Beavers 9

Short-term Regulation 18

The First Winter of Regulation 19

Inventory-taking 24

The Second Winter of Regulation 26

Beaver Studies 34

River Lodges 34

A Night with the Beavers 36

Beaver Life at the Outlying Stock Farm 42

Beavers at School 54

The Beaver Class 54

The First Lodge 58

The Territory 61

Digging Without Earth 64

Fina and Esmeralda 66

Joys and Disappointments 68

In Pastures Green 70

Winter: Under Ice and Under Roof 80

In the Beavers' Nursery 85

Bachelor Life 99

In the Artificial Beaver Stream 109

The End of the Story and the Beginning of Another One 123

Family Life 128

"Beaver Pedagogy" 139

Bibliography 151

Index 152

Introduction

Even in the Space Age we still know very little about mammals —those inhabitants of earth that include us humans and are closest to us. The beaver is no exception in spite of the fact that it has been an object of human interest to perhaps a higher degree than any other animal.

In the days when the beaver was still common over large parts of the earth it acquired a well-deserved fame as a master builder. The results of its work were clearly evident and well calculated to stir the imagination: felled trees perhaps several feet across, "houses" that might be more than ten feet high, and dams that gave rise to large lakes.

The master builder himself appeared very seldom. He spent the greater part of his time in his underground passages and holes and beneath the surface of the water. He worked at his building labor only through the dark autumn nights, and then he was extremely shy and cautious.

The beaver was a constant object of the huntsman's attention and one might have thought that the daily handling of beavers' bodies would have counteracted the mystery and the formation of myths about the big rodent. But the beaver has in fact the capacity to appeal to the imagination even when dead; a somewhat unwieldy body often weighing more than forty-five pounds with a beautiful mammal's head, hands with five mobile fingers, feet rather like a duck's, and, finally, a broad, flattened, scale-covered tail (Fig. 2).

It seems that mankind in the course of the ages has regarded the beaver in two different lights: either as a supernatural being—as do, for example, the North American Indians—or else as a being with almost human qualities. The various concepts of the popular imagination about the beaver long ago formed the basis of beaver literature and many of them, strangely enough, still live on in the folk-lore, even of today. But the first time the beaver appear in literature—in Hippocrates' writing, 400 B.C.—it is a practical aspect of the animal that is under consideration, namely the medicinal use of the beaver's castoreum.

Castoreum is a yellow substance with a strong but not unpleasant smell. It is formed in two pockets, each the size of a clenched fist, which lie beside the genital organs in both males and females (Fig. 1), and as it was one of the most mysterious natural products known to man, it was considered to have healing properties. Castoreum has been used as a miraculous medicine from time immemorial and is often referred to in both ancient and medieval literature, which also reflect primitive ideas about the beaver's way of life.

Even by the sixteenth century the beaver had been eradicated from the greater part of Europe, but when the colonizers of the New World made acquaintance with beaver country the old legends, mingled with the ideas of the Indians, were revived.

As late as the eighteenth century the most fantastic accounts of the beaver still appeared in scientific works. But by then there were also more critical writers. Samuel Hearne wrote ironically in 1771 that the only thing that remained to make the natural history of the beaver complete would be to draw up "a vocabulary of the beaver's language, a codification of its laws and an account of its religion."

In 1758 Linnaeus gave the European beaver the name of *Castor fiber*. It was not until 1820 that the American kind was given its scientific name: *Castor canadensis*.

Around the end of the nineteenth and the beginning of the twentieth centuries a number of realistic accounts of the beaver were written that still remain unsurpassed. They are based on more or less casual observations and on studies of the results of the beaver's work.

Modern biologists have not made any really important contribution. Of recent years American and Russian workers in particular have carried out a number of systematic studies in beaver country, but none of them has fully described the beaver's activities on land, under water and in his underground passages. Consequently the literature as a whole gives an incomplete and in many respects erroneous picture of the beaver.

Modern behavior research, or ethology, has opened up fresh possibilities for studying the ways of animals. One method used consists of rearing an animal from a tender age on a bottle, so that it builds up social contacts with its keeper, who then has the opportunity of studying its "inborn" behavior.

Since the summer of 1957 I have devoted a large part of my time to the beaver, and since the summer of 1959 my family and I have lived with beavers, which have regarded us as members of their group or as parents. In this book I shall try to tell a little about our experiences and discuss some of the objective observations and evaluations. I hope to be able to show that an animal does not lose its charm if one refrains from humanizing it. Many authors who have written popular books about animals have equipped them with at any rate some of the qualities of the human psyche, but I cannot understand why animals should be considered any more interesting or delightful for it.

Some aspects of an animal's behavior we shall perhaps never be able to describe in scientific terms. How, for example, can we ever be able to know what an animal "feels"? It looks as though animals have feelings that are related to our own, but we can, of course, only speculate. On the other hand it is often possible to show that the "feelings" one sometimes seems to recognize are not necessarily associated with intelligence in the strict sense of the word.

It was the catastrophe that occurred to the beavers of the river Faxälven that started me on my study. I owe a great debt of gratitude to the Faxälven beavers and to Elias Eriksson, the owner of the homestead at Ramsele. It was because of him that the animals were not left to die, and since that time he has helped me greatly with my work. But Elias' call for help might have sounded unanswered had not the provincial Gamekeepers' Association had such a far-sighted committee and Tall von Post as adviser.

This book would never have been written had we not had in Sweden just one representative of modern behavior study. Docent Eric Fabricius, who is head of the Ethological Department of the Zoological Institute at Stockholm University, has shown throughout a stimulating interest in the beaver studies, which were financially possible only because of his international contacts.

But first and foremost I have my wife to thank for enabling me to devote so much of my time to the beavers. Our boy announced at an early stage that he was "tired of beavers," but now he too is beginning to take an interest.

My Beaver Colony

A Little Beaver History

As my friend Elias Eriksson is always saying: "There's no animal so like humans as beavers." I hope the reader will soon come to realize that this statement is more flattering to mankind than Elias really means it to be.

There are actually a number of elements in the beaver's behavior that are apt to arouse a sympathetic smile in the observer, and there is no other animal that can by its labor transform the landscape in the same way as can the beaver and man.

In comparison with the beaver's tampering with nature, mankind's is a recent phenomenon. By the time certain ape-like beings were beginning to assume a resemblance to man, there had already been beavers for a long time and presumably they had then already mastered the art of dam building.

By now it is already so long since the beaver disappeared as an integral part of the landscape that we have completely forgotten what it once meant. If one wants to have any idea of what the world once looked like one must remember that it was inhabited by beavers. In ancient days they were numerous in the greater part of Europe, North America, and large areas of northern Asia. They built their dams in all the gently flowing smaller watercourses so that one dam ran into the next. They played their part in seeing to it that the countryside was well watered and that effluence through the rivers was even. The beavers' dams also gave rise to a change in the geology and topography of the smaller valleys, and along the riverbanks the beavers took care of harvesting and forestry, ensuring that vegetation followed a particular course.

The Beaver and Primitive Man

Wherever primitive man lived in close contact with the beaver it had a great, indeed often a fundamental, significance for man's economy and his philosophy.

The North American forest Indians based their economy on the hunting of beavers. Almost every part of the dead animal could be used. The flesh was regarded as a delicacy. The fat was used as a specific against frostbite, among other things, and the castoreum was regarded as an effective treatment for practically every physical ill. The skin was used to make clothes, rugs, moccasins, stockings, rope and many other articles. The dead were wrapped in beaverskin, and to give a skin to anyone as a present was a sign of particular friendship. The bones were made into tools of various kinds, and the chisel-shaped front teeth were the Indians' principal edge tool. Even some of the Indians' pleasures came from the beaver, for castoreum was considered to give tobacco a particularly pleasant taste and a beneficial effect. To the Indians the beaver was a holy animal, which often figured in their religious legends.

The various tribes and families had their own beaver grounds from which they took their tribute on a scale based on the practice of centuries. A well-calculated balance was preserved between the beaver and the beast of prey, man.

Since the spirit of the beavers was considered to be closely related to that of man and consequently able to influence it, the Indians always had to show respect to the beaver. Before hunting they offered prayers and promised that they would treat their victim well after death. After feasting on the flesh they laid the bones against an altar and threw all the remains that they could not use back into the water.

Among the various peoples that inhabited the beaver lands of Asia a form of social life arose that has been called by Russian experts "beaver economy," since it was based on beaver hunting.

These primitive peoples, too, regarded the beaver as a holy animal.

Discoveries at ancient dwelling sites, religious centers and graves indicate that the beaver occupied a similar position among the hunting tribes of Europe. One testimony to its importance in the philosophy of the early Europeans is the number of place names of which its various names form components, such as *biur* and *bäver* in Sweden. There are also, of course, all the legends about the beaver that have been recorded and which to a certain extent live on even today.

The Beaver and Civilization

As Western civilization spread, the beaver was driven out: in part because man took the lands previously occupied by the beaver for cultivation, and in part because civilized man had no thought of preserving natural resources. He just took it for granted that the abundance of nature could never cease. In any case he could never have dreamed it was possible to exterminate the beaver. There is said to be a proverb in Jämtland, Sweden, to the effect that, if everything else vanished, the hare would not vanish from the wood nor the beaver from the meadow. The economic interest of civilized man in the beaver was based at first on the animal's medicinal importance. The fur had always been highly valued, but it was above all the insatiable demand for castoreum that brought about the intensive persecution of the beaver in Europe.

In the greater part of the Mediterranean area the beaver disappeared early, with the leveling of the forests, but even at the beginning of our present era there were still beavers in Spain and along the River Po in Italy. Toward the end of the thirteenth century they were exterminated in England. In Sweden they began to diminish in large numbers during the sixteenth century, and in the course of the eighteenth century they disappeared from all the more cultivated parts of the country. The last beavers in the forests of Norrland were killed sometime around the end of the nineteenth century.

After that it seemed to be only a question of time before the European beaver would cease to exist at all. Only a few remnants of the species, none of the groups consisting of more than a few hundred animals, remained in southern Norway, on the middle stretch of the Elbe in Germany, in the Rhône delta in France, in Poland and in certain areas of Russia. But right up to the end of the nineteenth century there had been plenty of beaverskin and castoreum on the world market due to the large supply of beavers in America and Siberia.

The Great Beaver Slaughter

After the hunting civilizations had vanished from Europe it was a long time before mankind's interest in furs began to revive. About the middle of the sixteenth century attempts were made to establish a fur trade between Europe and the New World, but not until the beginning of the seventeenth century was this organized in a business-like way. After that, one station after another, mainly for handling beaver skins, was set up in French Canada, including one at Quebec in 1604 and at Montreal in 1611. This latter developed more rapidly and soon became the main center. A number of trading stations were founded by Christian missionaries who proved not averse to making a good income for themselves in their spare time by buying up beaver skins from the Indians.

The event that was perhaps to mean most for the development of the fur trade occurred in 1670 when the Hudson's Bay Company acquired control over large areas of land in return for its having undertaken the exploration of the Northwest Passage.

When the governor of French Canada tried to take over the Hudson's Bay Company, competition between the English and

French led to open war. As soon as conditions had grown a little more peaceful, the fur trade expanded rapidly. A single trading station could supply 20,000 beaver skins a year. Although exports included a number of valuable products, beaver skins represented about two-thirds of the whole value. They served for a long time as the only currency. Not merely other types of skins but other goods that were traded in were valued in terms of beaver skins. At one trading station in 1733, for example, a pair of shoes cost one beaver, while a rifle cost ten or twelve.

At that time the Hudson's Bay Company was exporting more than 200,000 beaver skins a year to Europe, and it has been estimated that about half a million beavers must have been killed every year in North America at the time when the hunting was at its height. The skins were sold at great auctions in London, where the hatmakers were the biggest customers. They made, from the fine underfur, the famous beaver hats.

At the end of the eighteenth century two new companies were formed, one being the North-West Company of Montreal, whose members seem to have thought of nothing but beavers. They founded an exclusive club, which was called the Beaver Club. The director lived at the Beaver Palace and the company minted its own beaver coinage. Shareholders drew enormous profits and lived like mercantile princes. The riches and power of the three main companies were reflected in the bloody strife that existed between them.

By the end of the nineteenth century the beaver had been exterminated over the greater part of the continent. The Indians had sold their daily bread and their souls for fire water and gimcracks.

By that time the last wild haunts in the Old World had also been emptied of beavers. In Russia, trade in beaver skins had been extensive up to the time of Czar Peter and exceeded in value that in all other types of fur. But once the beaver hunting in Siberia fell into the hands of Russian colonists, the beaver was soon exterminated there too.

By the end of the nineteenth century the beaver had finally

played out its part in world history. For long ages it had given the bare necessities of life to a great many people. For a few centuries it had provided luxury and power for a favored few who were ready to fight one another to ensure their share of the wealth. In the wild rush for beaver skins, the North American backwaters had been opened up to Western civilization. Where once stood simple fur-trading stations there are now huge cosmopolitan cities.

The Beaver Returns

Not until the twentieth century did the white man in America begin to realize that the exploitation of natural resources could not go on in the same thoughtless way as before.

Land erosion, for example, threatened to undermine American agriculture. When the prairies were cultivated and the forests in the adjoining mountain tracts felled, the level of groundwater sank. The soil had been cultivated without a thought for future generations, and its capacity to absorb and retain water was reduced. Since large areas lay periodically without a protective covering of vegetation, the dry earth, fine as dust, could easily blow away. People's eyes began to be opened to the fact that the beaver had played a part in maintaining the water supply in these areas. The protection and reintroduction of the beaver became a matter of national urgency and was among the intensive measures taken by the authorities to check land erosion. Effective methods of capturing and redistributing beavers were considered; experiments were even tried in which they were dropped from aircraft.

The results of these efforts were not long in showing themselves. Since the 1930s the American beaver has been increasing in numbers and now covers a large part of its original distribution area. Certain tracts are even considered to have too many beavers in relation to the present resources of the countryside. When food gives out the beaver abandons its dams, and then, of course, its favorable effects soon cease. The American preservation authorities are now concentrating on maintaining a balance between the number of beavers and the rate of growth of the beavers' food in order to keep the beaver grounds continually inhabited.

In Europe the Russians have shown the greatest interest in the reintroduction of the beaver. They have done a great deal of work on experiments in breeding beavers and setting them loose in the wilds. The beaver is now to be found again in large parts of the European Soviet Union, and in recent years there have been increased efforts to encourage the still dwindling stocks in Siberia.

That one little remnant of the original Scandinavian stock has been rescued in Norway is due to the fact that one man in the 1840s protected the beavers on his own lands. Since protection has been extended to cover the whole country the stock has increased to such an extent that, at the beginning of the twentieth century, it was estimated to number several hundred animals. Since then the Norwegians have been in a position to supply beavers for Sweden and Finland.

The first pair of Norwegian beavers was set free in Sweden in 1922 in Jämtland. From 1922 until 1939 about a hundred beavers were imported and released in various places, from Småland in the south to Norrbotten in the north. On the whole these attempts at reintroduction gave poor results, which is not surprising considering how little was known about the biology of the beaver. People were not even sure how to determine the sex of the animals, so that it was by no means certain that a pair consisted of a male and a female. Really vigorous colonies have been established only in Värmland-Dalecarlia and Jämtland-Ångermanland. There are now about two thousand beavers in Sweden.

Norwegian beavers were introduced into Finland in 1935. A couple of years later a few pairs of American beavers were also taken there. By 1955 the stock had grown to 400 to 500 animals, three-quarters of which were believed to be of American origin. It would appear that the American beaver increases more rapidly than the European. It is believed that American and European beavers crossbreed out in the open and that the cross product can in turn propagate itself. The Russians, too, have introduced a number of American beavers, so that there is a serious danger that our European beaver will gradually become only a part of the American animal.

One need no longer have any fears for the beaver's existence. Of course it has not the same opportunities now as it had in the original countryside, but there will doubtless always be large areas in which the beaver can flourish. The American beaver has good prospects in Canada and large parts of the U.S.A. For the European beaver the forest areas of Norrland would seen to offer great possibilities.

Elias and the Beavers

Elias Eriksson lives on the top of a steep sandy riverbank with a view over Homsele between Vagnforsen and Rabbstuguforsen. In 1957, when I met him for the first time, it was one of the most beautiful places I had ever seen. Tourists on their way up through the Ström valley often stopped on the main road opposite Elias' farm to admire the view over Faxälven.

Elias' principal occupation is forestry, but like so many other Norrlanders he is also an avid hunter.

When the beavers came to Ramsele at the beginning of the 1940s the shooting of hares and forest birds was almost finished and the supply of fish in the rivers was running low. The word "shooting" to a Norrlander is practically synonymous with elk shooting, but to a really enthusiastic nature lover the increase in the number of elk was no proper compensation for the loss of the fish in the river and a teeming population of hares and birds in the forest. So that the news that there were beavers on Holmeselet was received with joy by all nature lovers in the vicinity.

Many people who live in close contact with nature are familiar with the old signs about weather and wind and other natural phenomena. Whether these are reliable or not is another matter, but by paying constant attention to the changes of nature some clearheaded people develop a sharpness of observation that to many townspeople often appears almost superhuman. Elias has a particularly highly developed capacity for registering natural phenomena, which he interprets in the light of his experiences over many years of closeness with nature, while at the same time he attaches great importance to the old signs.

At first it was only Elias who saw that there were beavers on

Holmeselet. For the last few years of the 1930s and the first of the 1940s he never saw a live beaver, but traces of various kinds were proof enough. When he was ottering on the river in the late summer evenings he often heard a loud splash out in the main channel of the river, but although he always drew his otter over the spot he never caught any big fish.

Elias did not know a great deal about beavers at that time. He had heard his father say that there used to be beavers in his grandfather's time and that in those days they cured most sicknesses with castoreum. He had also heard how the beavers felled trees and built lodges, and that the male used the female as a sledge to transport timber. Not far from the house where Elias grew up there is an island called Beaver House Island. That name stirred his imagination when he was a little boy. He used to picture the beaver houses as square, timbered buildings with several floors.

One autumn Elias discovered that the beavers were living under a steep sandy bank. The entrance to their home was deep under the surface of the water. A year or so later he caught sight of them for the first time. They were sitting there at the edge of the water under the bank one evening when he came down to the river, but they slipped silently out into the water as soon as they saw him. Now he knew that it was possible to see the beavers and after that he saw them swimming in the river every summer's evening he had occasion to go there. This also solved the mystery of the "big fish" that kept splashing. It was the beavers, striking the flat underside of their tails against the surface of the water when they were frightened and dived.

Elias began to examine carefully those parts of the bank where the beavers went ashore. He grew more and more familiar with their habits and soon he was able to see beavers on land in the summertime any evening or morning he wished. After that he went down to the river more and more often, simply to see the beavers.

The animals had fixed habits. Elias could almost set his watch by them. In the evenings they swam around in the water for

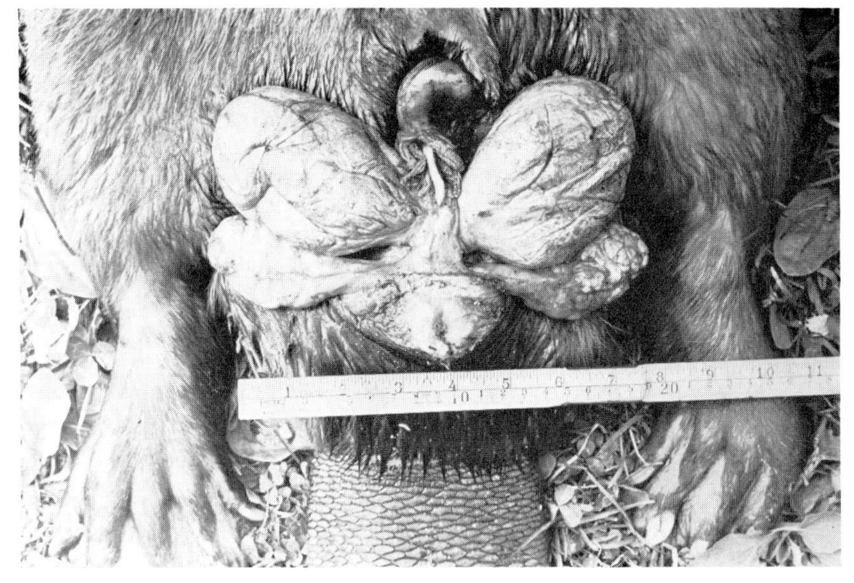

1 The two castoreum pouches, the size of a clenched fist, and the oil glands underneath them open into the "cloaca," which in this photograph has been turned inside out. The penis bone in the genital passage, which has been cut open, shows that the animal is a male.

2 The tail is covered with horny plates, which replace the greatly reduced covering of hair.

3 Branches and trunks which are not more than about four and a half inches thick are cut into suitable lengths and dragged the shortest way down to the water.

4 As soon as the beavers are able to get up on land in the early spring they start felling trees again for a short period.

5 Land ice has covered over the beavers' winter store on Faxälven at Gammelmo near Ramsele.

6 Two young beavers born that year in the three-foot-deep roadway. At this age they do not normally go so far from the water.

7 Beaver dam more than six feet high. The downstream side consists of branches that lie in the main parallel with the direction of the current, so that they form a support for the dam. *Photo Lars Björk.*

8 The first snow covers the river lodge in the foreground. The winter store, which stretches out some yards into the ten-foot-deep water outside the lodge, is solid from the bottom up to the surface of the water. Before the ice is formed it may stretch out perhaps fifteen yards into the river.

9 From the space between the ice and the water the beavers have dug a tunnel underneath the snow to a felling-place in the wood.

10 A stream lodge more than six feet high in the autumn. The beavers have just covered it with a layer of fresh mud and barked sticks.

11 Thanks to the pool of water inside the feeding chamber the beavers are able to sit and eat at the water's edge even in the wintertime. The pieces of wood are rotated between their hands as the bark is stripped off.

12 Ice lies as a protecting cover over the beaver dam and the winter store in front of the stream lodge.

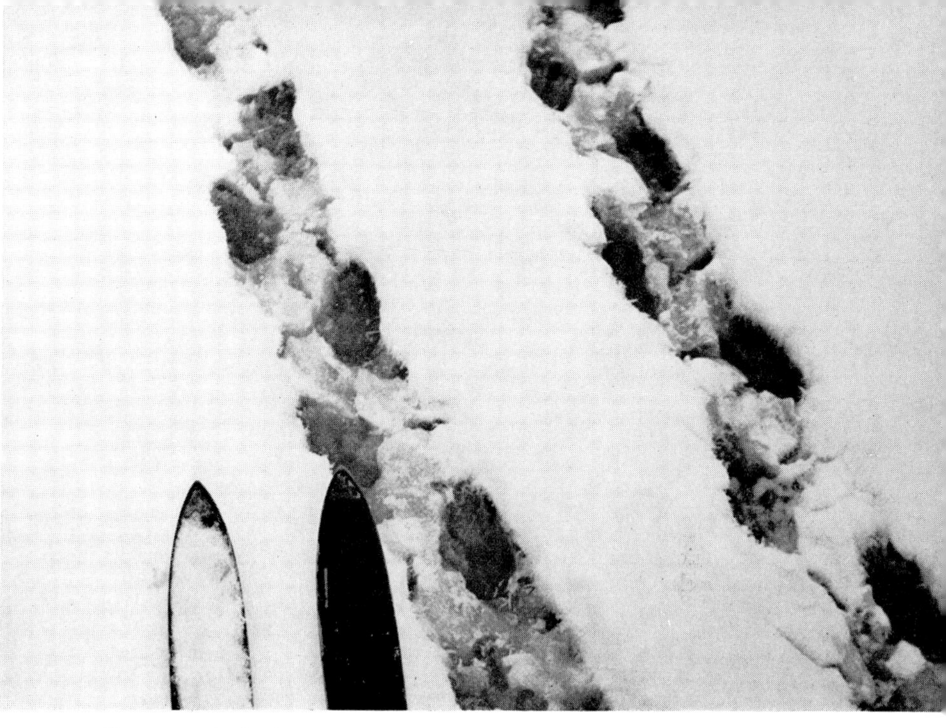

13 Beaver tracks in the snow.

14 Beavers prefer to sit out of doors to eat, if the temperature is not lower than about twenty-four degrees.

15 Young beavers are very fond of apples. The one in the photograph is about a month old. Notice the mobile fingers of the hand, the webbing between the toes and the double claw on the second toe.

a while before going up on land at fixed spots. From each such landing place a beaver path led into the riverside vegetation. In summertime the beavers dragged small leafy branches or whole armfuls of plants down to the bank and sat there eating just at the edge of the water, at pre-established feeding places. Both paths and feeding places little by little became well worn and were a noticeable feature of the riverbanks. Any kind of foliage would apparently do, but in choosing plants the beavers were more particular. Their very favorite dish for a great part of the summer was apparently rose bay.

When the leaves of trees began to turn yellow it became increasingly common to see twigs and branches stripped of their bark. They might be branches of any deciduous tree that grew in the district, except alder.

Occasionally Elias managed to get so close to the beavers while they were sitting eating that he could see how they held the pieces of wood in their hands while they stripped off the bark.

One autumn toward the end of the 1940s the beavers built their first proper lodge on Holmeselet a little way upstream from the place where they had first settled. They suddenly started collecting branches of alder, stripped sticks and other wood that was of no use for eating and piled it up on the shore from the edge of the water up to the steep bank. It appeared that the beavers had passages beneath the level of the shore into the bank just underneath the pile they had collected. Of course Elias never got a glimpse of the beavers during the dark autumn evenings and autumn nights, but in the daytime he was able to check what they had done at night.

They worked night after night until the pile had assumed sizable proportions. Then they started covering it over with a layer of bark residue that had dropped from floated timber, which they fished up from the bottom of the river. They kept on at this for a long time until all unevennesses had been smoothed out. Then they laid an even layer of fine twigs over the bark; finally they plastered clay all over it. The woodpile was thus converted into a firm, well-insulated roof, which covered the ground over the passages through which they passed from the bottom of the river into the bank. Elias realized that this must be a beaver lodge, though he had never seen one before.

While the work on the lodge was going on the beavers were busy felling trees in the woods along the riverside. Elias had already noticed that they started felling larger trees when they changed over from their summer diet to their autumn diet of bark. They were evidently particularly fond of aspen bark, for they tracked down odd aspen trees several hundred yards from the river. The size of the tree made no difference. There was an enormous birch tree on Holmeselet, more than three feet in diameter at its base. The beavers started on it once or twice, as they often do even with smaller trees, but one autumn they went at it in earnest and then it was not long before it was felled.

The felled trunks lay in all directions, as they had fallen, but the trees that grew closest to the river often fell out into the water, because they leaned that way or had better developed crowns on that side. Often a falling tree was held fast by another. The beavers sometimes gnawed through the trunk again, a little higher up, but that seldom produced any better results.

After a tree had fallen, the animals gnawed off one branch after another and dragged them one by one the shortest route to the water (Fig. 3). The paths by which they dragged them grew into deep furrows, from which all obstructions were gnawed or dug away, so that the roadway was absolutely smooth. If the bank was steep, the roadway was dug down through it at a slight angle to the water's edge, which sometimes meant digging furrows several feet deep with walls more than knee-high between the water's edge and the flatter ground above (Fig. 6). Sometimes they dug a tunnel instead, from the top of the bank down to water level. As all forest workers know, the organization of suitable transport lines is a very important detail.

Trunks and branches that were more than about five inches in diameter were left where the tree was felled. The beavers often stripped them there, but just as often they left them unstripped.

Some of the wood that was transported down to the water was cut into smaller pieces, which were stripped at the feeding places, but after the animals had eaten all they wanted they went on dragging down branches that they floated down to their home. They gathered them together in a pile in the water in front of the entrance.

This was the most remarkable thing of all. In front of the newly built lodge the water was almost ten feet deep and the current was strong. First the beavers stuck the edible branches into the part of the lodge that reaches out a little way into the water. Then they wove them together and built on in this manner from the bottom up to the surface and from the lodge farther out into the water (Fig. 8). The marvel was how the beavers managed to get the branches to stay where they were in the deep, turbulent water and how they could weave them together into such a compact mass. When winter came and put an end to wood-gathering, it was quite safe to walk out on the woodpile about fifty feet straight out into the river. The pile was five to ten feet high from the bottom of the water up to the surface, about ten feet wide, and contained approximately 3,000 cubic feet of closely packed wood.

The great woodpile in the water puzzled the villagers considerably. One thought that the beavers were building a dam right across Faxälven, another that they were going to build a bridge across to the other side. Elias watched carefully to see what would happen next.

On Holmeselet the river was usually open until November or December. It took severe cold—temperatures of at least twenty degrees—to make the water freeze over. The later the ice formed, the less cold was required for anything dramatic to happen. Ice started to form on land even in October (Fig. 5), and, if the ice formed late, it could grow out from both banks until it almost met in the middle of the river.

The freezing over of the river always began with the formation of an ice dam. Quite suddenly ice would appear on all the rocks that stuck up out of the water. The bed of the river became as it were glazed over with ice. At the head of the fall of Vagnforsen the glazing grew thicker and thicker so that the water began to rise. The land ice was lifted up by the rising water, which flowed more and more sluggishly, partly because of the damming up, partly because the water was full of blocks of ice. When the water had risen more than three feet, the top few inches suddenly stopped moving and turned into a layer of ice. Due to this, the temperature of the water rose, so that the glazing of ice on the river bed and the head of the force melted and the water level slowly began to sink under the covering of ice, which a few days afterward dropped down with a crash. The newly formed ice made contact with the surface of the water, but the land ice, which had had time to grow thicker, and which, with the freezing water raised up around it, had frozen hard onto the bank, formed a ledge against the bank a good distance above the surface of the water.

Elias had watched this process many winters. This winter it would be interesting to see what happened to the branches the beavers had gathered together in the water. By October the pile was covered over with land ice, which was soon so thick that it effectively cut the beavers off from the bank. It seemed a little alarming when all traces of them suddenly vanished and an impenetrable layer of ice and frozen earth formed over their dwelling place. How could such large animals survive for six months, shut away in their underground passages in the river-bank?

Twigs sticking up out of the land ice here and there marked where the beavers' woodpile stood. The top twelve inches of the pile froze into the ice and the rest was down in the water. Strangely enough, the beavers had laid alder on top, just in that part of the pile that froze into the water.

When the water dammed up, just before the ice formed over it, the brushwood pile rose with the land ice when this was lifted up by the rising water. When the ice sank back after the covering had formed, the outer part of the pile made contact again with the bed of the river, while the part that lay closer to the lodge hung free underneath the ice. Just beside the lodge an air space formed between the land ice and the water.

Toward the end of the winter season the ice began to melt over the brushwood pile, so that soon it lay in open water. All the wood that had been frozen in the ice lay floating in the water, but all the rest was gone. Some of the pieces of wood floating in the open water bore traces of beavers' teeth, and it was not hard to imagine what had happened to the rest. Now Elias knew that the brushwood pile had been the beavers' food supply for the winter season, and it looked as though they had been far-seeing enough to build the part of the store that was going to freeze into the ice of alder wood, which was of no use for food.

After the ice had thawed there were pieces of wood stripped of their bark scattered along the bank for a good distance on either side of the beaver lodge. The beavers had obviously had plenty of space during the winter. Before the ice came they had been shut in in their system of passages. When they needed food they dived down through the underwater end of the system of passages and out through the exit down by the bed of the river, and, having trimmed off suitable pieces from their winter supply under the ice covering, they hastened back to the air-filled parts of the passage system. But after the ice had formed over the water the beavers could also go to the space between the land ice and the bank to sit and strip the bark from the sticks from their winter store.

How, then, could the beavers get enough air in the passages, which opened out only underneath the water and which were completely covered with an airtight roof? Of course they would bring a good deal of air in with them when they returned from the water, but animals that weigh more than forty-five pounds must use up a good deal of oxygen, and as far as Elias could judge five beavers must have spent the winter in that lodge. Elias solved this problem during the following winters.

He noticed that the beavers never plastered their lodges with mud right up against the side of the bank. Hoarfrost always formed there on cold winter days, and if it was really severe cold there was sometimes a sort of plume of steam rising from the upper part of the lodge. This obviously acted as a ventilator. If the bank was very steep and high one could find places where warm air percolated out high up in the bank. The actual dwelling holes obviously lay deep inside the bank, and in steep banks that were not frozen over so heavily the passages passed quite close to the ground. But in general the passages and living holes were covered by a thick layer of frozen earth and the upper part of the brushwood roof served as the only air intake.

The thaw usually came in April on Holmeselet and as soon as the beavers could get up on land they started felling trees again (Fig. 4). They were often so eager that they felled more trees than they actually needed at the time. As soon as the leaves began to appear they changed over to a diet of leaves and plants, and then the branches of the trees they had felled in the spring were left untouched.

In the spring there were young beavers that were not even half grown. They would have been born the preceding spring or summer, but Elias never saw young beavers in the summer. He knew nothing about the beavers and their young. Everything that happened inside the beavers' passages and holes was still a well-kept secret.

In the first half of the 1950s a number of beaver lodges appeared on Holmeselet. Every autumn the old lodges that were still inhabited were repaired with alternate layers of branches, twigs and mud and bark residue from the bed of the river. Because of the big winter stores it was easy to see in the autumn which lodges were to serve as wintering places for a group of beavers. By the middle of the 1950s three lodges were inhabited at the same time and Elias estimated the number of beavers on Holmeselet at about fifteen.

The beavers quickly spread along the river. Before long there was a toal of six inhabited beaver lodges in the Ramsele community. Up by the old church one could sometimes see as many as fifteen beavers swimming around at once. Reports that beavers had settled at various places along the river came with increasing frequency. At a farm some twenty miles downstream from Ramsele one could sit at the kitchen window and watch the beavers swimming about in the river in front of the lodge.

Faxälven valley was really a place for watching beavers. They were, after all, the nearest neighbors to the people of the valley, and through the light summer nights one could sit on the high riverbanks and watch the beavers without their being in the least disturbed.

Short-term Regulation

The valley of Faxälven is a very beautiful valley. All along the twenty-five mile stretch between Ramsele and Lake Helgum the river winds between steep banks. Well-built villages lie on fertile terraces above these banks, which are often clothed with woods of birch and aspen. Plant life is luxuriant on the old hay fields along the riverside and on the islands in the river. On the south banks the lily of the valley is often out as early as in southern Sweden, and in the summer everything is blue with wild geraniums and harebells, white with ox-eye daisies and cow parsley, and yellow with buttercups and marsh marigolds. Ramsele has its own specialty, the riverbank anemone, and lady's slipper grows in profusion close to the river.

It was a pleasant valley for the beavers. In the summertime they were able to eat the plants and foliage that grew just beside their lodges. The deciduous woods were dense and they had them to themselves. For human beings it was hard to drag the timber up the steep banks home to the village, but to the beavers, which lived down by the water, the deciduous woods were easily accessible. Thanks to the fertile soil, the trees grew quickly again, particularly on the side of the valley exposed to the sun. It would have taken a good many beavers to keep the woods properly thinned.

It was very easy for the beavers to dig passages in the fine sediment of the high banks, which were usually so high that the dwelling holes could be hollowed out above the highest water line. In the lower part of the shore the beavers' homes might become submerged in the high spring floods, but they have learned through millions of years experience to cope with situations like that. On account of the land ice forming early and the thaw coming late the beavers were unable to get up onto dry land for five or six months of the year. Due, however, to the good supply of deciduous trees quite close to their lodges, they had time to collect ample winter stores, and once the ice had formed they had plenty of well-protected space underneath the land ice.

The First Winter of Regulation

At the beginning of December 1956 Elias was busy driving timber out onto the ice of Holmeselet. He had just heard that the river was to be placed under short-term regulation. No one in the village knew what that meant, but they had never had any unpleasant experiences with water regulation. When the second large power station on Faxälven was completed, a long-term regulation was introduced, which meant that the seasonal variations in the height of the water were evened out. The spring floods were modified and this was of a certain advantage both to beavers and humans. On the other hand the formation of ice was somewhat delayed, so that the beavers had to wait longer for the extra space under the ice, but in the main their conditions on Faxälven remained unchanged after the introduction of long-term regulation. The term short-term regulation sounded, if anything, less objectionable than long-term regulation, and Elias could not imagine that anything was going to happen that would appreciably affect either himself or the beavers.

One Sunday about the middle of November he had taken a walk along Holmeselet to see how the beavers were getting along with their winter preparations. All seemed well. It had been a mild autumn and the animals had had time to collect a large food store beside the three lodges. For the second half of November and the first week of December the weather had been normal for that time of year, with temperatures of fourteen to five degrees, so that the winter stores had been well anchored in the land ice. As far as could be humanly judged, the beavers could feel secure.

On December 10 the water rose about twelve inches in the course of the morning. This was nothing very extraordinary, but after that the variations in the height of the water grew greater every day. Toward the end of December the water was low in the morning, but by midday the river had risen about thirty-two inches. With the rising water came quantities of ice. The thick land ice along the entire river was broken loose by the rising water and slid down into Holmeselet. To this was added the ice that formed on the low water every night. The river was full of great ice floes of up to 100,000 square feet, often with wooded clumps of shore attached.

Elias had never seen anything like it, though he had lived beside the river all of his life. He began to wonder how things would go for the beavers.

The day before Christmas Eve he went down to the beaver lodges, where the horror of desolation confronted him. Both the land ice and their winter stores were gone. Thanks to the mild weather the beavers had been able to get up on land after the land ice had been carried away, and great fellings showed that they had been in feverish activity; but what they collected during the nights had been washed away by the water when it rose in the mornings.

During the Christmas holidays there was to be no short-term regulation, and if it turned cold the river would freeze over so that the beavers would be shut in without the possibility of getting food.

Elias' mind was not easy as he walked home from the river. Since the beavers were protected he would be summonsed if he did anything that affected their buildings. Then could a public undertaking destroy their winter stores all along the river with impunity? Elias was not allowed even to maltreat his own horse. Then had a public undertaking the right to starve the beavers slowly to death? Or was cruelty to animals and the killing of protected animals allowed if it was done on a big enough scale and if the people concerned made enough money out of it?

Elias decided to apply to the Game Preservation authorities to try to get help. He had always been a little shy of authorities, but he asked a neighbor who was much more articulate than he to call the Game Preservation adviser, Tall von Post, in Härnösand. Tall von Post was ill, but he himself had wondered what effect the short-term regulation might have on the beavers, and he promised to see that the matter be investigated after the holiday.

Chrismas came, the river froze over and the beavers were shut into their passages. On the day after New Year's Day regulation was resumed. The ice was broken up, but as the severe cold continued the ice floes remained. On Twelfth Night the weather turned milder and it snowed. The ice floes began to move and patches of open water appeared here and there.

The beavers were still alive in spite of their long fast and they made use of the openings in the ice to go up on land after food. They trampled out a path in the sixteen-inch-deep snow to a clearing eighty yards from the river.

Since the weather was mild Elias was able to lay out food on the shore above the high-water line. The beavers took it the very first night, but the mild weather did not last long and the variations in the height of the water grew greater and greater. The land ice that was formed in the night was carried away by the rising water during the day, and the food he had laid out often went with the ice into the river.

Ridges of ice formed along the shore, and grew every day to about high-water level, so that they came to be more than three feet high, with a broad overhang out toward the water. In the daytime the water stood on a level with the upper part of the ice, but at night, when the beavers were active, it was more than three feet from the surface of the water to the overhanging projection, so that the beavers could not reach their feeding places on the shore. Elias then tried laying down large trees, with their crowns in the water and their root ends up on the ridge of ice.

The idea was that the beavers eat off the branches from the top. During milder periods all went well, but when it was cold the trees froze fast in the ice. Later Elias tried tying large bundles of branches together so that they sank into the water completely.

One way and another he managed to keep the beavers supplied with food, but it was possible only because the winter was unusually mild; he was worried about what would happen if there was a long period of severe cold.

He wondered also how the beavers farther downstream were getting on. He had read in the newspaper that they had felled trees at a spot where, unfortunately, it was possible for the owner of the land to collect the timber. Eight loads of wood that the beavers had felled and lopped he had driven home in his sledge. Neither the landowner nor the newspaper reporter seemed to realize why the beavers were felling trees so feverishly in the middle of the winter.

In February, Elias tried to organize a more permanent winter store for the beavers on the Holmeselet. Two wagonloads of brushwood were sunk into the water in front of each of the lodges and tethered with wire rope to firm anchorages on dry land. The question was whether the ropes would hold during holidays, when the water regulation was stopped. If severe cold occurred then, heavy ice would form on land only to break away from the shore in blocks of perhaps 100,000 square feet. Anchoring a block that size with wire rope was, of course, out of the question.

The ropes held, however, right up to the beginning of March. Then began the coldest part of the winter. The wire ropes that had long been under a hard strain snapped and the ice floes outside the beaver lodges started off down the river with all that remained of the beavers' winter stores. The cold grew worse and worse. Now the beavers were again dependent on day-to-day help. They fetched some of the food that was laid out for them on the newly formed land ice, but the cold was so severe that they could remain in the open for only brief periods. Elias had noticed in the course of the years that beavers never go outside if the temperature is below the low twenties.

On the morning of March 21 a beaver had been up in the snow at one of the feeding places. The food, which had been laid out the previous day, had frozen into the ice and the beavers had obviously been without food all night. Elias stood absolutely still to see if the beaver would come back. After a minute or two it climbed up onto the land ice and started burrowing for food in the six-inch-deep snow. It seemed thin and feeble and its backbone showed clearly. This was proof that the beavers were in want and were dependent on help.

After that Elias saw beavers almost every time he went to the river to lay out food. Hunger caused them to lose some of their shyness and sometimes they even came out while he was cutting wood for them and dragging it up. They made desperate attempts to crawl up over the ridges of ice, and when they had to give up they would start eating fir trees that had been carried out into the river by the ice.

Toward the end of March the cold began to ease and the noonday thaw ate away at the land ice, but it was to be a long time yet before the overhanging ice had been sufficiently reduced for the beavers to get up onto the shore. Elias hacked paths for them through the barricades of ice and carefully sanded them, and the beavers at once made use of them to get up to the woods on the shore. Little by little spring came. Elias was able to write in the diary he had kept carefully all winter: "So this, for the beavers, terrible and uncertain winter is over."

On April 12 Elias had his most remarkable meeting so far with a beaver. He was walking along the shore and when he was about ten feet away from a beaver path he saw a beaver coming in from the wood on its way to the water. It did not stop until it was fifteen feet from him. He and the beaver stood staring at one another for five minutes, after which the beaver went calmly on and passed within ten feet of him without showing the slightest fear. Not even when it got out into the water did it hurry away.

Inventory-taking

In the spring of 1957 I went up to Ångermanland. I had been asked to take an inventory of the beaver groups along Faxälven and to try to find out how they were being affected by the river regulation. Tall von Post met me at Härnösand and went on with me to Ramsele. There I met Elias for the first time. He had a pleasant, outlying stock farm about three miles from home, and there I was allowed to establish myself for the next few summers.

The work was to begin with an inventory-taking journey from Ramsele to Lake Helgum. Tall had planned the journey before I arrived and we were able to set out the following morning from Ramsele in two row boats. The journey took three days, and I shall never forget it.

We rowed slowly under the high, steep banks, carefully studying the shores. There was no difficulty in establishing where the beavers were. Most of them, of course, were asleep inside the banks, but every now and then one of them actually bobbed up in front of the boats, even though it was broad daylight. At night there was obviously plenty going on on shore. We rowed past one feeding place after the other where stalks and leafy twigs floating at the edge of the water showed what the beavers had been eating the night before. Here and there we could see their well-trampled paths into the vegetation or the woods. The deeply excavated roadways through which they had dragged the wood for their winter stores the previous autumn had been well worn. Sometimes we could see from the boat how these roads ran from the edge of the water up to the crest of almost perpendicular banks some one hundred feet high. At the felling places trunks lay scattered in every direction. The lodges, or more correctly "roofs," at the points on the shore where the passage systems ran down to the bottom of the river at times assumed considerable dimensions. One was fifty or sixty feet in diameter and filled up the whole space from the water's edge to the top of the six-foot-high shore. One could see through the clear water how the beavers had hollowed out deep furrows in the bottom up to the entrances to their passages.

Close to most of the lodges there was a hole in the bank above the water line, and tracks in the sand showed that they were inhabited by beavers. At some places we found entrances to passages without a "roof" at a good distance from any lodge, but there there were no roadways in the vicinity. It seemed as though only passage systems with a "roof" over the entrance were inhabited during the autumn and winter.

We marked in the beaver lodges and entrances to passages on a map and made notes of all indications of their presence. Roadways, felling places, summer paths and feeding places were seen on the shores both upstream and downstream from the lodges. At their thickest the territories of the different colonies seemed to run into one another without any noticeable boundary.

Along that twenty-five mile stretch of river we found twenty-eight lodges, which had certainly been inhabited the previous autumn. We were very impressed by the intensive activity the beavers had shown. The summer tracks showed that there were still plenty of beavers along the river, but there were also signs indicating that they had had a difficult time in the winter. When we passed through Ramsele, where Elias knew all the beavers' lodges, we found that a couple of them were gone, and the winter fellings at several places farther downstream showed that the beavers had lost their winter stores and had been forced to go up after food in the middle of winter.

This journey had given me one of my finest experiences of nature. It was sad to think that this unique addition to the Swedish countryside was possibly doomed to disappear. If one reckons with five beavers to every lodge, that would mean that there were 140 beavers on Faxälven, and that would probably have been only a good beginning if the river had escaped the grim fate of being put under short-term regulation. In America they reckon six beaver lodges to the mile as normal if conditions are really good, and I found it hard to imagine better conditions for beavers than were offered by the greater part of the stretch of river between Ramsele and Lake Helgum. Assuming that twenty miles of this stretch presented the best type of beaver ground, it should normally have been able to support about six hundred beavers.

What Elias had told me and what I had been able to see for myself had given me many problems to ponder over, but the summer passed quickly. I watched the beavers when they were active outside their lodges on Faxälven at night, and by day I investigated the abandoned beaver lodges and other products of their work. I also began an inventory of the beavers on the tributary rivers. On my journeys inside the Faxälven basin I was able to observe the attitude of the people toward the beavers, and in my wanderings along the beaver streams I learned to know the beaver landscape.

When I went home I still knew little enough about the beaver, and my head was full of problems. One thing was certain: the beaver fascinated me. I would go on with the work and try to solve some of the problems.

The Second Winter of Regulation

Elias and Tall had promised to keep me informed of what happened on Faxälven the following winter.

During the Christmas holiday, 1957, I had a letter from Elias. It was more like an S.O.S.

Later I was sent his notes for the period up to Christmas, and during March and April I received one letter after the other. Together they make a thick bundle, and they give a detailed account of conditions on Holmeselet during the winter of 1957–58.

That autumn the beavers changed over as usual to a bark diet and began to fell large trees in the middle of August. They also started repairing their lodges. They even worked early in the afternoon, so that once Elias was able to watch a beaver pressing on mud with its nose from a distance of only fifteen feet.

The short-term regulation had been discontinued during the summer and was resumed at the beginning of September. The lodge that had been most affected by the regulation the previous winter had been abandoned, but the animals that had lived there were trying to build a new home on a part of the shore where the variations in water level were particularly acute. Elias measured them every day with a tide gauge. When they reached three feet, the beavers had to abandon their building. When winter came they would be homeless and would have no hope of survival.

One night in the middle of September the beavers started collecting stores in front of the two remaining lodges, but it was a hopeless undertaking since the variations in water level increased toward the end of September to more than four feet. In the mornings the beavers' stores lay high above the surface of the water, and later in the day they were floating loose in the deep, fast-running water. One day the stores sailed away from both the lodges.

The beavers went on collecting with incredible energy. By the beginning of October they had again collected quite large stores, but one day one store departed and a few days later the other. Although the animals had now lost two large stores they went on working. It was not hard to foresee that sooner or later they would see their work laid waste for the third time, and Elias thought very much about how he could possibly help them.

After long consideration he got a neighbor to help him. Together they drove down poles into the bottom of the river to form a fenced-in area the size of a winter store in front of each lodge. The beavers were not disturbed by the poles. As long as they held, the wood the beavers collected was safe, but the stores grew larger and heavier and the variations in water level increased day by day.

In the end one of the stores, which stretched out nearly twenty feet into the river and was so compact that Elias could walk right out on it, started slowly drifting with the current. In spite of desperate attempts to anchor it to the shore with heavy iron hooks, the poles were soon torn out on the downstream side, after which the greater part of the store drifted away with the current.

Elias was nearly desperate. He began to wonder whether to blow up the beaver lodges with dynamite to save the animals suffering, but Tall refused to hear of it. A week later he came to make an inventory along the river.

He did the whole stretch down to Lake Helgum by motorboat in two days. It was a far less pleasant trip than the one we had done the summer before. It was cold and bleak on the river and the beaver territories looked desolate. Most of the lodges had not been repaired and several had collapsed. Only at a few were there winter stores of normal size. By several of the lodges scrapings on the river bed showed that the stores had been swept away. Long stretches where the beavers had excellent feeding grounds and where the summer before there had been many well-maintained lodges now lay quite desolate.

Up to the beginning of December the beavers on Holmeselet managed reasonably well in spite of everything. Due to the regulation, the river was open all the time. The weather was unusually mild, the ice formed slowly along the shores and so the beavers were able to work on indefatigably at their stores. By the lodge where they had already lost two winter stores the third grew rapidly, but at the other lodge, where they had already lost three large stores, things were not going so well. And that was not to be wondered at. The collecting was a major effort. Trees had to be felled, branches dragged a distance of fifty or a hundred yards down to the water, and then floated on to the store, where they had to be fastened carefully beneath the surface of the water. And it was no small quantity with which they were dealing. Each one of the stores that had floated away contained more wood than could be loaded onto a large truck. The woods closest to the lodge were beginning to give out and the transport lines grew longer and longer. And besides, the spells of intense cold, when the beavers could not be out, were growing longer.

Then in December winter began in earnest, with temperatures of twenty degrees below zero and lower. By day land ice formed and froze fast to the shores during high water, but as soon as the weather turned milder it would break up and be swept away by the strong current. During the day the high water stood four feet nine inches above the low water at night, and the high-water level soon rose farther because of an ice dam of a kind Elias had never seen before. But as long as the land ice remained there was no danger, and so he tried to anchor it just around the lodges.

About the middle of the month it turned mild again. The land ice broke up and floated away. It remained outside the lodges due to it having been anchored, but a few days later the ropes gave way and the ice, with the stores, sailed away. Although the beavers had worked intensively the whole autumn and collected three and four stores of normal size, they were left without food when winter began.

Elias concluded his report for the period up to Christmas with the following lines: "The fact is, that as regards the large colony of fully protected beavers on Faxälven, those responsible for the regulating of the river have been guilty of cruelty to animals on a scale probably unequalled in this country. To judge by what has happened already the presence of beavers along this river will very soon be no more than a memory. To say that all this was necessary for the so-called public good is incorrect. The 'public good' would have been no less well served if the river had been cleared of beavers *in a humane way* before the water regulating began."

After Christmas the regulation became more severe than ever. The twenty-four-hour variations in water level soon amounted to more than six feet. The land ice that had formed over the Christmas holiday was broken up and great ice floes began to pile up on Holmeselet so that the water was dammed up to about ten feet above its normal level. Timber-floaters who were well acquainted with the river told Elias that it was the same both upstream and downstream from Ramsele. Tall went up to look at the devastation Elias had told him about over the telephone. The water was almost up to the ventilation hole of one of the lodges and Elias was afraid the river would rise still farther and that the animals would be forced out of their lodge. When they got there two beavers were swimming uneasily about in the water outside their lodge.

The following day it was colder. The water had risen during the night and ice had formed over the ventilation hole. When Elias went out onto the newly formed ice, he felt something soft under his foot. It was a dead beaver. It had obviously been forced out of its lodge and been trapped in the broken ice while trying in vain to get air. Presumably all the inhabitants of the lodge had perished.

Elias felt almost a sense of triumph while at the same time he found the whole thing extremely distasteful. Now at least he would be able to prove how terribly contrary to nature all this short-term regulation was, so that it would be possible to put an end to the whole miserable thing. At that time he still believed that some sort of consideration would be shown to the people and animals who lived in the Faxälven valley. A few days later the lodge was completely destroyed by the violent variations in water level.

There was a danger that more ice would pile up and that the other lodge, too, would be swamped, and one day Elias saw from home through his field glasses that the water had risen still farther. He went down to the river. The temperature was six to twelve degrees below zero. The beavers had been driven out of their lodge by the high water. They were sitting on the extreme edge of the land ice, dipping their feet alternately into the water so as not to freeze fast. The water had turned to ice on their coats, which were so dishevelled that the animals were wet to the skin. It would be seven hours before the water fell enough to enable them to get inside the lodge and warm themselves, and they would presumably freeze to death before that. The animals no longer looked like beavers, sitting there shivering with ruffled coats that gave not the least protection against the intense cold. This was cruelty to animals in the very highest degree, and Elias could not stand and look at it any longer but went to fetch his elk rifle. On the way back to the river he had second thoughts. Obviously the only proper thing would be to put an end to the beavers' suffering, but he had begun to think of the consequences. He would probably be prosecuted and lose his gun license. He stayed indoors for the rest of the day so he wouldn't have to see what happened to the beavers on the edge of the ice.

Toward the end of January the beaver lodges along the whole river were inspected once again. All but five were seriously threatened. The ice on the shore was high, in some places up to ten feet and with a broad overhang. It would be almost spring before it melted and the beavers could get up on land.

At the beginning of February it turned milder and the water level fell somewhat. Toward the end of the winter more ice floes had collected, so that they covered the entire river beside the lodge in which, in spite of everything, a few animals were still alive. Elias hacked a hole in the ice in front of the lodge and filled it with brushwood to protect the entrances to the passages against broken ice forcing its way in. He had come to realize that the beavers needed their winter stores outside the entrances to the passages, not merely as food, but also as protection against broken ice.

There was not very much left of winter, and it was possible to save the remaining six animals. As soon as it was feasible for them to get up on shore, Elias cut gaps for them through the ridges of ice so that they could go up to the woods and get food for themselves.

After relating all this in a letter dated March 16, 1958, Elias concludes: "Why should such a state of things be allowed to continue in what we call a civilized land. It is enough to make one skeptical of the world-wide storm of indignation at the fate of the space dog Laika in Sputnik II."

I had written during the winter to the River Regulation Board and asked for information regarding the estimated variations in the height of the water, etc. Later I received an invitation to visit the head office in Stockholm. The information they had available for me was of no use at all. They had probably had a good laugh at my clumsy attempts to get some sort of information, and made no bones about the fact that they really had no idea how great the variations in water level might be with the twenty-four-hour regulation they were applying.

The prospects of co-operation with the water regulation people seemed not very bright. I realized that the only thing I could do was to gather together facts about the beavers' living requirements and the possibilities of its surmounting the unnatural conditions imposed by short-term regulation.

After the second winter of regulation it seemed pretty clear that the beavers had no future on Faxälven. In that case, I thought, one ought to try to capture such animals as remained and move them to safer waters. The following summer the Ramsele power station would be in operation. The outlet tunnel ran out downstream from Vagnforsen, and Holmeselet would, accordingly, be laid dry. The first thing was to catch the beavers there.

Tall had already made a succession of unsuccessful attempts in the early spring to catch them with nets and in other ways. Later in the spring he was busy taking films on Holmeselet. He was making a documentary about the beaver catastrophe. Elias gave him what help he could, as he was anxious that the Swedish people should have a chance to see what had happened. One day he saw a beaver in a little pool of water. He hurried over to it. As the beaver had no possibility of getting out of the water it pressed itself down against the bottom, motionless. It only needed to stick its nose out of the water for a brief moment every fifteen minutes and it was not much to film. But this was their chance to catch a beaver. There was a wild hunt in the pool and in the end they managed to get him into a net. Elias was very upset by it all. The wet, frightened beaver in the net looked very different from the beavers he had been accustomed to seeing in undisturbed activity on Holmeselet. The captured beaver was to be kept in captivity until the two other members of the family had been caught, and then they were all to be set free at the same spot. With the aid of nets and a sort of fish-trap arrangement they did finally succeed in capturing the others, but by this time the first beaver was dead.

Then summer came again and I was able to resume my studies on Faxälven. The first thing I saw was that the beautiful stretch of Holmeselet had been reduced to an indescribably ugly and stony desert. The beaver lodges were stranded high up on dry land. In the deepest part of the river channel there was still a certain amount of water flowing, and there were deeper residues of water here and there. Two beavers still remained, and they stayed, in spite of everything, all through the summer. In August, Elias got a neighbor, Alfons Eriksson, to help catch them. This was relatively easy as by now the river was practically dry, but after that two beavers appeared that Elias had never seen before all through the summer. They started building a lodge beside one of the deepest hollows and they went on even after the Regulation Board had started cleaning out the river bed. One day the great bulldozers started excavating the part of the shore where the lodge stood, but Elias talked to the foreman and arranged for the beavers to be left in peace. Both animals survived the following winter.

The Regulation Board had been instructed to build impounding reservoirs on Holmeselet. Elias thought that these would provide a suitable environment for both fish and beavers, and was glad that the two beavers were so attached to Holmeselet that they had stayed, in spite of having roaring machines on their doorstep.

Beaver Studies

During the winters I spent all my spare time going through all the literature I could find about beavers. It was easy to make contact with American experts, who gave me helpful advice and sent books and reprints of articles. The Russians, too, were helpful, although it took a long time to get their letters and articles translated.

I grew more and more astonished at how little was known about the beaver. The abandoned lodges provided an opportunity of finding out what the beaver's living quarters were really like, and it was a very exciting moment the first time I opened up a beaver lodge. Since then I have investigated a great many lodges of different types along Faxälven and the wooded streams.

River Lodges

Along Faxälven all the beavers lived during the winter in lodges of the type I have already described. Since these are characteristic of larger watercourses we normally call them river lodges, but they can also be found beside streams, provided the banks are steep and consist of material into which it is easy for the beavers to dig passages.

River lodges are generally built beside deep water, so the beavers can start digging into the shore far below the surface of the water. There are usually at least two passages that lead from the river bed diagonally upward into the ground underneath the "roof" of branches, fallen bark, twigs and the like. Just where the passages into the ground reach up above water level they widen into a chamber, the floor of which is almost level

16 Once the babies had learned to drink from the bottle the difficulty was to stop them before they had had too much.

17 A certain time after they have stopped eating in the morning, beavers sit up on their tails and noisily lap up the green, porridge-like substance that forms in the appendix.

18 Small babies that are not being looked after by a female get soaked to the skin when they bathe. These in the photograph are washing energetically after a bath in a little pool in the dry river bed of Faxälven.

19-20 The double claw on the second toe of the hind foot acts like a pair of pincers when beavers comb the fine under-fur of their coat. The toe with the double claw is bent, so that it stands almost at right angles to the other toes, and the two arms of the pincers close tightly on the fine hairs as the toe is drawn upwards and downwards through the fur.

21 Four two-month-old beavers eating on the shore of Faxälven.

22-23 Grooming beside the pool of water inside the feeding chamber of the lodge.

24 Three of the youngsters eating beside the pool in the school terrarium.

25 The beaver's right hand. Branches and pieces of wood are gripped between the "little finger" and the other fingers. The "thumb" (farthest to the right in the photograph) is very small.

26 The youngsters have started at the age of about four months building a lodge around their box in the farthest corner of the terrarium.

27 Young beaver walking on his hind legs while carrying a few pieces of wood in his mouth and an armful of shredded wood, sticks and twigs held tight between arms and chin.

28 In Zürich Zoo, Professor Hediger's American female beaver has had young several times. She carries them in the same way as our beavers carry building material.
Photo Werner Haller.

with the surface of the water and contains a little pool. We knew that beavers always like to sit and eat just at the edge of the water, and with the pool in the feeding chamber they can sit at the very edge of the water and bark sticks from their winter store, even during that period of the winter when there is no space they can inhabit between the ice and the shore.

The feeding chamber is obviously a very important feature, since it occurs in all beaver lodges. After I had found that out, I began to picture what life was like in the lodges during the winter. I wished many times that I had had a TV camera inside the feeding chamber so that I could have had a direct view of the beavers' winter life. In the spring perhaps I should even have been able to see the female beaver bathing her young in the pool!

From the feeding chamber one or more passages lead farther into the bank. They pass close under that part of the "roof" that lies farthest in to land and which has not been sealed with bark residue and mud, but after passing the "roof" they go deep into the bank. They are very often long and are often forked. The height is ten to fourteen inches and the width twelve to sixteen inches. Deep inside the bank are one or more sleeping chambers high above the highest high-water level, where the beavers sleep on beds of "shredded wood."

When one knows what it is like inside the lodges it is not hard to understand why the beavers take so much care over their "roof." A well-built "roof" is an effective protection for the passages where they pass quite close to the ground in the part of the shore most likely to erode. During the winters of regulation it often happened that the "roof" was torn away by ice, and the part of the passages that were filled with water and the feeding chamber were then eroded away by ice and water. In lodges that had lost their "roofs" during the winter the beavers had little chance of surviving. Because the "roof" is so well insulated the water does not freeze in those parts of the passages that are under water or in the pool in the feeding chamber. And since the back parts of the "roof" are not plastered over with finer material, ventilation is assured even in those parts of the passage system that lie deep under the surface of the ground.

It is not necessary to have a space between the ice and the water since the beavers have a well-organized feeding place inside the lodge, but the increased space is certainly welcome when the water level sinks under the ice cover later on in the winter.

The construction of the lodges is adapted to a more or less constant winter water level. It is not such a serious matter if they are temporarily flooded during the spring, when the beavers can seek emergency quarters in some temporary hole, but if the lodges are flooded during the winter the situation can spell catastrophe.

Of course the appearance of the lodges varies considerably according to the conformation of the shore, but all entrances to the passage system always lie under water and there seems to be always a feeding chamber with a pool under the "roof" and an air intake at the back of it. Sometimes there is more than one feeding chamber under the same "roof," with a common passage system or with two separate ones.

I assumed at first that the beavers always begin by excavating the passages from the water inward under the shore, and then build the "roof" over them, but in some lodges both the feeding chamber and the passages down to the water were hollowed out from the actual mass of branches and other materials forming the "roof." The beavers had obviously been forced by the nature of the shore to build the "roof" first and then make passages between the water and the bank through the material they had collected.

A Night with the Beavers

When one stood before an opened beaver lodge one often wondered what it had been like when the beavers were there. How many animals had lived there? What had their family relationships been like? What happened when their young were born? But I saw no possibility at that time of ever finding out. The only means available of studying beavers were to sit and watch them through the summer nights when they were outside their lodges.

We spent a week at the beginning of July on the top of a high bank. From this vantage point we could see about half a mile of the river and three beaver lodges. They had all been inhabited during the winter and in the one farthest upstream there was a pair of adult beavers and a young one all through the summer. In the two other lodges, which were quite close to one another almost half a mile downstream, there were several adults and several young animals, but we could not make out which of the two lodges the various animals came from.

Around the upper lodge there were a number of beaver paths that all ended in a well-trodden spot up on the bank, which was quite low. We could not understand why the beavers had been up there. The grass was worn away and the trampled earth was black. The signs indicated that the beavers did not go there to eat or to collect food, although there were unusually few feeding places at the water's edge close to the lodge. Between the two other lodges feeding places were, on the contrary, unusually frequent, but there were no beaver paths of this mysterious kind there.

The first evening at the lookout post was fine and warm. When it turned colder toward morning we were able to light a fire to warm ourselves without disturbing the beavers. We could keep them under constant surveillance through our field glasses as long as they were outside their lodges and we could see quite well what they were doing.

The first beaver of the evening bobbed up in the water in front of the farther lodge about half past seven. It climbed up onto the "roof," where it sat and appeared to take its ease in the evening sun, its fur gleaming reddish-brown. After a while it began its morning, or rather evening, toilet. It sat with its tail forward and its strong thighs spread out, rubbing its plump stomach long and carefully with its hands. Then it stretched up and scratched itself on the back. It combed its arms carefully one by one and fully scrubbed its thighs with both hands. The whole thing looked almost human and was delightful to watch. I had never even read about the beaver's toilet ritual and was quite fascinated by the remarkable sight.

One beaver after another came out from the lodges. They all swam around for a time before going ashore. At the upper lodge one after another climbed up the paths to the mysterious places on the riverbank, where they executed a series of extraordinary movements. They scratched with their hands, then humped their backs and strained, at the same time kicking one back leg in the air. I had seen some of these movements before and knew that they meant that the beavers were depositing castoreum. But previously I had only seen them do it on small piles of earth that they had first scratched together close to the water. One often saw these heaps along the shores close to the lodges and they always smelled strongly and pungently, but not unpleasantly, of castoreum.

In the books I had read I had discovered two different theories as to the significance of castoreum. According to one the secretion was a form of enticement between the sexes, according to the other it was used to mark out territory. Now perhaps I should get some insight into the beavers' behavior in connection with depositing castoreum and perhaps be able to work out what it meant.

After the three beavers had each visited all the castoreum places around the lodge they set off downstream. They swam a good distance and went ashore between the two lower lodges, where the animals from these two were already busy searching for food.

The shore on the other side of the river was now full of life. A long row of large dark lumps was sitting there at the edge of the water. Every now and then one of them vanished into the shore wood and sometimes two would meet at a feeding place or out in the water. From time to time a young tree would begin to sway, and then it was not long before a beaver would appear, energetically dragging a leafy bough down to the water.

The animals used the different feeding places interchangeably and without any apparent order. There seemed never to be any hostility between animals from different lodges. They all acted quite independently in getting food and eating, but animals from the same lodge did seem to hold together in a way. Now and then two adults would meet at a feeding place and sit and groom one another, or go out into the water where they would splash about together for a while before returning each to its own feeding place. One one-year-old was often fussed over by a larger animal which smoothed out the young one's fur with its teeth.

After midnight three of the beavers left the feeding places and started swimming upstream. They went quite slowly. We knew what it was like from rowing against the current, and we admired the way they made use of all the back eddies and other peculiarities of the current that might make it easier for them.

When the three animals reached the vicinity of the lodge farthest upstream, where they evidently lived, they went ashore. They nosed around for a while at various spots on the shore and then all of a sudden broke into intensive activity down by the river, walking restlessly up and down the steep bank and straining and kicking vigorously at the castoreum spots.

The mist lay heavy a little above the water. The atmosphere of night helped to give one a sensation of something primeval when the big heavy animals clambered awkwardly up the bank and carried out their strange ritual up there on the crown. In the end they all assembled out in the water where they swam around for a while together before returning to their feeding places.

The depositing of castoreum was obviously something very significant since it seemed to be the only reason why the animals swam that long way against the current and back again. Possibly they were inspecting their territory and making sure that no strangers had penetrated inside while the owners were away.

Possibly fresh castoreum on the castoreum spots frightened strangers away from landing on the shore by the lodge. The animals from the lower lodges did not seem to be so particular about inspecting and marking out the land around their lodges, but then they remained in the vicinity all the time.

I could not understand why the animals from the upper lodge had necessarily to swim down to the lower lodges to eat at night. Admittedly the supplies of summer food were best there, but they would have been able to manage well enough with what grew close to their own home.

Toward morning an osprey came sweeping down over the shore. All the beavers rushed headlong into the water and those that were already there dived, striking their tails against the surface of the water so that the sound could be heard a long way down the river. Later that summer I had plenty of opportunity to study the way beavers react to disturbances of this sort.

Our observation post offered almost unique opportunities for studying the beavers, but the light summer nights do not last long. Just as the beavers' behavior began to be really interesting the dark autumn nights rendered all observation impossible. I realized that it would be many years before I could collect sufficient observations to be able to draw any definite conclusions about the beavers' family relationships and territorial behavior. And probably the beavers would have vanished from Faxälven before then.

During the winter of 1958 I thought a great deal about what I should do in order to learn more about beavers. Occasionally the idea crossed my mind that I should try working with tame animals, but it was difficult to convince myself that such a thing would be feasible.

The following autumn term I had other things to think about than beavers. But a long series of incidents in the course of the autumn suddenly transformed a vague dream into realistic plans. I obtained an adequate grant and got ready to start experiments with beaver the following summer.

After Christmas we moved to Sollefteå. At the school there is a terrarium room measuring twenty-five by fifteen feet, with tiled walls and a cement water tank. The headmaster complained that this expensive piece of equipment had received very little use. I promised that I would use it, but I did not reveal then what I had in mind.

My main task was to catch some beavers, as young as possible, and bring them up on a bottle. I would study their development to try to find out what was inborn in a beaver's behavior and what it normally had to learn from its parents or from its own experience. I had also promised Tall to help in catching as many of the Faxälven beavers as possible. I had made contact with the wild life protection people in America, who had experience in capturing and moving beavers and we had ordered American traps. At that stage, of course, we were quite convinced beavers no longer had any future on Faxälven.

Beaver Life at the Outlying Stock Farm

As I have said before, we did not know when female beavers give birth to their young. I had seen statements in literature on beavers to the effect that in Russia and on the Elbe it was usually at the end of May or the beginning of June. To be on the safe side I waited until June 23 to set out from Ramsele.

Eric and Pian Fabricius had come up from Stockholm and Tall from Härnösand. Alfons came with us as "beaver catcher," since Elias was skeptical about the whole business and did not want to have anything to do with catching beavers ever again.

The previous night I had inspected several beaver lodges in the wood. At one I had seen an old beaver diving with a green twig in its mouth and so I felt convinced that there were young beavers in that lodge.

When we got there at last with all our equipment we began by closing the stream both upstream and downstream with nets. Tall and I, each armed with a large sack, then placed ourselves at the entrances while Alfons set about the lodge with spade, crowbar and saw. As it was a long wait we soon laid aside our sacks to help with the work. After several hours had passed, we got down to a passage filled with water and deep underneath the compact mass of branches, twigs and earth. Then Alfons started digging at the back of the lodge.

Suddenly something splashed into the water in the opened passage. I dived down with my hand, and when I brought it out I had a young beaver by the tail. It was the size of a large rat and looked wet and miserable, but it soon dried out and revealed itself as a delightful apparition. round and soft, with a funny flat handle behind. It showed no trace of fear. Shyness is an attribute that is simply not developed in beavers as young as that.

We found no more young ones and we were very surprised that we had not seen a sign of the adult animals. It was only later that I realized how cautious beavers are in their lodges and how they can slip away completely unseen when they are disturbed.

It was high time to be thinking about feeding our protégé. I had read of Indian women in the old days bringing up young beavers at the breast and I assumed that European beavers too would be able to manage on the food one gives young babies. The stores cupboard back at the farm was equipped with a packet of dried milk for this purpose. But I suddenly remembered on the way home that I had forgotten to buy a feeding bottle. When we got back to the village the shops were all closed, but I knocked at the back door of the drugstore and was met with complete understanding. It was, of course, written all over me that I was a happy new father.

I was really very satisfied with my day. The first necessity, if I was to carry through my plan, was to get hold of young beavers. Most of the people I had talked to about it had shaken their heads. Now I knew that it was possible. And as yet I knew nothing about the difficulties I was going to encounter in rearing them.

My troubles began when we got home to the farm. Before I could even think about my own screaming stomach I had to prepare the baby's bottle. This I did in accordance with all the rules of the game, but even so, and even though the youngster must have been very hungry, it pushed the strange "teat" contemptuously away with one hand. I had to force it into the little one's mouth, but what milk ran in was soon blown out at the corners. Every four hours the whole night through I tried again, with the same poor results.

The next day we repeated the procedure of the day before at another stream lodge and there we got two young ones.

Now we had three and that had to do, at any rate for the present. When we got back to the farm we let all three of them loose outside, and soon they were hard at it investigating their surroundings. In the end they started digging with their hands in the moist earth. It was one of the first of the beaver's movement patterns I had been able to see at close quarters. A few days later I also saw them shoveling away the loosened earth with their hind paws, which make excellent shovels, as they are webbed for swimming.

All through the night Tall and I took turns feeding the babies every four hours, but for all our trouble we got only about ten grams into them each time. Number One showed an inclination to try and nibble a sallow leaf.

The next night I was alone with the beavers in one of the rooms at the farm, where I let them loose on the floor. Before long they were sitting each in his corner, noisily sucking at sallow leaves. I couldn't see that they got anything from them, but it was a good sign anyway. After a while they started roaming restlessly around on the floor, and whenever they happened to meet they gave a faint piping sound, so high-pitched that I could barely hear it. In the course of the night they investigated every corner of the room. The following day they slept in a box, but in the evening I let them out on the floor again. By now they obviously felt at home, for they showed none of the uneasy wanderlust of the previous night and each sat in its corner, gnawing at a willow branch or trying to eat leaves, which they handled very clumsily.

Later that night I was witness to a spectacle that was to me more fascinating and exciting than almost anything I had read or seen before. The little balls of wool that had previously seemed so shy and reserved moved out onto the floor quickly and energetically and suddenly each one of them began to behave as though it were the sole owner of the room. When they came anywhere near each other, one would often start dancing, that is to say flinging its head from side to side and then flinging its body sideways or turning around. At the height of the dance it would rise up on its back legs and fence at another one with its arms. It was an indescribable sight to see these little miniature beavers standing up on two legs supported by their funny little flat tails, measuring blows at each other with solemn mien. After a few rounds the boxing gave way to wrestling. They gripped each other firmly by the skin and pushed for all they were worth. Now one of them would manage to push his opponent a few steps back, then the other would get the temporary ascendant. Only after they had pushed one another backward and forward for perhaps ten minutes would one of them give way and the match would be over for the time being.

Although the wrestling was often so fierce as to produce sharp cries, they never bit one another and did not appear to be really aggressive.

After a night that had been strenuous both for them and for me, they fell asleep in a heap behind the back door. They always sat there eating together the following nights. After that I never saw them wrestling again, though they quarreled over their food like good brothers and sisters.

Much later I realized that the wrestling had been an expression of the little ones' territorial behavior, which was aroused as soon as they began to get to know their new surroundings, but which was still so weakly developed in the baby beavers that it could be worked off by means of these quite innocent wrestling matches. After one night's confrontations Number One and brother and sister Numbers Two and Three had got to know one another and formed a group.

On the third day we prepared an enclosure with a little dam outside the house, where the youngsters would be able to bathe for a little while every day. From the first moment they plunged into the water they were transformed. As soon as we made a sudden movement they dived, striking their tails against the water so that it splashed in every direction. They behaved exactly like beavers out in the open and swam both above and under water with fully developed swimming movements, but they soon calmed down and sat for a long time grooming themselves on the shore. The complicated grooming movements were also fully developed and I enjoyed seeing them at such close quarters. After they had been in the enclosure a few times they were tame there, too.

After three days of captivity they started sucking energetically as soon as they got the "teat" in their mouths. Before long they were stretching their hands out eagerly for the bottle as soon as they saw it, and chewing at it hard as they fed. After that, the difficulty was not to let them have too much (Fig. 16). They all weighed about three and a quarter pounds and we estimated their stomach capacity at 50 milliliters and decided they should not have more than 50 grams of milk at a time. They soon fell into a regular rhythm under which they slept all day and were active all night.

Although they had begun to eat quite a quantity of aspen leaves, we had never seen any excreta, and were getting uneasy that they might have some sort of stoppage. One evening we put a bath of water on the floor. They bathed constantly that night and in the morning the water was thick as soup. It was obvious that beavers have to have access to water to be able to fulfill their wants. After that they bathed regularly in the bath, groomed themselves carefully afterward, and then sat down to eat at the feeding place. They were extremely clean and pleasant to have about the house.

One of them had begun to "talk" to me. It often came up and nuzzled at me and made a long succession of whimpering sounds, so highly nuanced that it really sounded as though it were talking. It looked at me all the time with what seemed to be an expression of confidence, and it felt quite strange to experience such a personal contact with an animal I had previously had such difficulty in learning anything about.

After a time the two others started talking, too. When they were out in the dam they were more reserved, but one evening one of them swam up to me to talk, and then I knew that the youngsters were really tame and regarded me as a member of their group. My shoes, funnily enough, had a special attraction for them and it could be quite troublesome, having the beavers clinging to my feet all the time. Sometimes I had to take my shoes off and put them in a corner, where they could clamber about on them as much as they liked.

The youngsters were so delightful that I thought it was really time I introduced them to the family. So that there should be room for us all I moved over to the other one of the two cabins that made up the farm, which was a little larger. Even so, it was quite crowded in the evening in the only room of the cabin, with three humans and three young beavers.

We went to bed at the same time as the beavers were getting out of their box and starting to bustle around, unabashed, on the floor. My wife, Ulla, and my son, Erik, looked a little astonished, but they agreed that the new members of the family were delightful. They had behaved in an exemplary manner so far, and I was sure they would make a good first impression. It never occurred to me that I had moved them to completely strange surroundings while they were asleep during the day. When they came out in the evening they set about exploring their new territory exhaustively.

We were awakened once or twice in the night by the beavers trying to climb up into our beds. After midnight Ulla woke me. She seemed very upset. As I came to, I realized all was not as it should be. The bath of water was empty. The youngsters had been bathing so much without grooming themselves afterward that they had slopped all the water onto the floor. In the ashes in the open fireplace sat one of the beavers. Ulla declared she had taken it out at least ten times, and she was obviously not exaggerating very much. The floor was covered with a mixture of soot and water, and the curtains were black as high as the beavers could reach. Our clothes were strewn all over the floor, impregnated with sooty water. The youngsters had obviously been amusing themselves by dragging them around on the floor.

In the morning the atmosphere was a little tense. Ulla wanted to know whether we were supposed to go on like this all through the summer. We scrubbed the room out and I tried to comfort and encourage her as well as I could. I realized that the coming night would decide whether or not the beavers were going to be able to get themselves accepted by the family.

Fortunately all went well. The youngsters had got to know their new territory and returned to their regular habits. They fed well now from the bottle, and since they ate a good deal of bark and leaves during the night we gave them milk only during the daytime. They were soon completely tame with Ulla and Erik, though they were on the alert and very reserved as soon as strangers came into the cottage.

It was completely idyllic and we all enjoyed the backwood life. A gray flycatcher hatched out its young between a couple of timbers in the wall, and a wagtail did the same under the roof just by the door. In the evenings young hares scampered about on the grazing land outside the window, and woodcocks brooded over their second covey; we could hear them even indoors. Erik, who was then seven, had a great deal to discover, and each day revealed to me some new and fascinating detail of the beavers' behavior.

But soon fresh troubles began. There were times when the youngsters were apathetic and lost their appetites. Our own spirits then went down too. We were already greatly attached to our protégés, and the rest of the work depended on how well we succeeded in rearing them. Normally they had such ravenous appetites that the difficulty was to break off in time before they had had too much out of the bottle. If they had too much milk they got diarrhea, and then the room smelt just as it would if a baby had an upset stomach. Normally young beavers have no smell that can be detected by humans.

I discovered that the youngsters ate evacuations direct out of the anal opening in the mornings (Fig. 17). At first I thought this had something to do with their stomachs being upset, but I found out later that it is a normal phenomenon, which occurs regularly with all beavers. Beavers have a very large appendix in which the cellulose is broken down by bacteria. A soft mass is then formed, which is rich in vitamin B_1 and which emerges a little while after the animals have gone to rest. They then sit on their back legs, put their noses down and lick up the green paste noisily, pressing around the anal opening with their hands. Excrement of the ordinary type is firm and dry and is passed only when the animal is in the water. Eating the evacuation from the appendix is a form of chewing the cud, which is a vital necessity for beavers, and seems to occur in some form in all rodents.

The young beaver that had grown tame first was particularly eager when he got the bottle, and one day he managed to swallow a whole 90 grams of milk before we could wrest the bottle from him. This happened twice in succession. The following day his bowels were loose and he did not want any milk at all. He seemed listless, and one morning he could hardly walk and in the end lay down on his side in a patch of sunshine on the floor. We realized then that it was serious and got into the car to take him to a vet, but the little one died in my lap on the way.

This was a hard blow. After that we became morbidly anxious at anything that might seem to suggest that anything could be wrong with the two others. But we need not have worried ourselves. They grew quickly and developed noticeably from week to week.

We used to close the kitchen door to stop the young beavers getting in our way there. They didn't like this. They knew there was something behind the door, but had not had an opportunity to explore it properly. The one we had caught first was the most active and had been particularly pigheaded and obstinate. It would scratch at the door for hours on end, until in the end we gave way and opened it.

The youngsters were now eating quite a lot of leaves and a good deal of bark and they handled the leaves and the sticks with increasing skill. They would hold the leaves with one hand while they bit off the leaf stalk at the base. After that they would roll up the leaf and hold it in both hands, pointing it obliquely forward while they fed it into their mouths. They took to biting off the twigs from the branch more and more often and then holding them between their hands while they barked them.

At the beginning of July the American beaver traps came. The Gamekeepers Association had appointed Alfons as catcher. I wanted to try to keep the captured animals until a whole group had been formed, so that all the members of the group could be liberated together at the same place.

The traps consist of two shanks covered with wire netting. They fall apart when the trap is set, so that it lies flat on the bottom when placed in shallow water. When a beaver comes swimming up to go ashore, the shanks come together and the beaver finds itself inside a sort of cage. The instructions accompanying the traps told one to fix a stick on either side of the trap to guide the animal toward the release mechanism.

The first week Alfons and I watched the traps continuously from the top of a high bank. They had been carefully placed, each in front of a feeding place. To eliminate our traces we had been very careful to pour water over all the places on shore where we had had to set our feet as we set the traps from a boat.

The first beaver that appeared was making straight for one of the traps, but when it got to twenty or thirty yards from the shore it stopped dead and lay in the water a long time watching the sticks, which were the only part of the mechanism that could be seen from above the water. Then it swam to another feeding place where there was no trap, but it came back about every half hour the whole night through to have another look at the sticks. Only after several nights did the beavers dare to go ashore at feeding places where there were sticks.

Since then I have come across many instances of beavers knowing their territory so well that they notice even the most insignificant changes. After that we set the traps without sticks and then the beavers usually swam straight in to the feeding places. We caught three in the first week.

They settled down quite happily in an enclosure, but later

we had to put animals from different groups together, and then there was trouble in the camp. It might be a week before they started eating again after we had put a stranger into the enclosure. The captured animals were always a source of anxiety, as for some time after being caught they were completely apathetic. It is not particularly stimulating to see newly captured wild animals.

The tame young beavers, on the other hand, seemed perfectly content with their existence. They were plump and smooth-coated and finer than any other beavers I had seen. Now and then we took them with us to the river or to a beaver stream in the woods where we let them swim about freely. After bathing for a time they always came back to wash themselves. They looked like little trolls sitting there on the shore combing their stomachs, arms, backs and thighs with their hands (Fig. 18).

When they were first caught they never got really wet when they bathed, and their fur was soft and dry almost as soon as they had got out of the water and shaken themselves. But after a few days of captivity they got wet through to the skin when they bathed, and had to work at their fur for a long time before they looked like beavers again. I did not know then how carefully the female beavers have to comb their young to keep their fur properly waterproof.

It was not until toward the end of the summer that the tame young ones learned to keep their fur in trim. By then the combing movements were obviously fully efficient, and the youngsters groomed one another increasingly often too. If one of them wanted his back groomed he would dance in front of one of the others and then start grooming the other one, which would respond immediately by doing the same. The two of them might go on grooming each other for a long time, always treating the corresponding place on the other's back.

Beavers have a double claw on the second toe of the hind foot, which acts as a pair of pincers when they groom the fine underpelt (Figs. 19 and 20). Our young ones were already using the double claw when they were first caught. Both combing with the double claw and grooming each other could be done at any time of the day or night, but the young ones cleaned themselves with their hands only when their fur was wet.

One day at the very end of July I moved the youngsters over to Elias' carpenter's shop. I had to leave Ramsele for a little while and Elias had promised to look after them until school began. He had been charmed by them when he came to see us at the farm during the summer and he was only too pleased to take care of them.

When I came back, fourteen days later, they had grown so much that I hardly knew them. They had not forgotten me; they came up and talked to me exactly as before. Elias was still giving them milk. They came rushing up as soon as their eyes fell on the bottle, and then they would stand on their hind legs chewing furiously at the "teat" as they sucked. In addition they were eating large quantities of leaves and bark, so we decided we could drop the feeding bottles for good.

The beavers had a large bath, which held upward of a hundred gallons of water, in which every night they performed a remarkable feat. They were now dropping quantities of excrement, which sank and formed a thick layer on the bottom of the bath, but by morning they had always gathered it all up and piled it on the edge of the bath, more than three feet high. As we did not then know the beavers' technique of carrying material by the armful, we could not imagine how they did it.

One night they managed to pull the plug out of the bath and all the water came out. When we went in in the morning they were splashing about in the water on the floor and seemed to be enjoying themselves enormously. After that, they pulled the plug out every night until, after many failures, we managed at last to outwit them.

I lived with the young beavers in the carpenter's shop several days and nights. They were quite strapping young creatures now to have poking around on the floor all night. Everything they did, they did with great energy. Number One was often to be seen carrying sticks around and he was liable to dump them down anywhere in the room. Later on they began to gather them together into a heap in front of their box.

At the end of August two more young ones were caught, and were gradually accepted as members of the group by Number One and Number Three. Our group consisted after that of four young ones, but the two caught last never became really tame (Fig. 21).

At the beginning of September the tame beavers started building in front of the entrance to their box, so that they had to make a fresh opening every morning, which they carefully filled up with shredded wood after going in to sleep in the mornings. They made a heap of all their refuse on the floor in front of their box, and farther out on the floor they collected all their surplus food. It looked as though they were trying to build a lodge with winter stores on the floor of the carpenter's shop.

Beavers at School

The Beaver Class

The summer holidays were over. The little group of beavers was to have its own classroom at the school. There was a large terrarium room on the fourth floor, which belonged to the Biology Institute, and in this we built a platform on a level with the top of a cement pool, which runs along one of the shorter walls of the room. Close to the other short wall we built a ramp down to the floor, and then divided off the whole area from the other half of the room with a partition. The platform was about eight by sixteen feet, and the area of floor on which we put the beavers' living box was about six feet in diameter. The terrarium looked very fine, with its tiled walls and teak edging at the windows.

The institute also boasts a small room with a glass wall onto the terrarium. We soundproofed the glass wall and set up a camera in front of a small peep-hole so that I could sit and watch the beavers and film them without their having any idea that I was there. The only question was whether they would appreciate their new home. The terrarium was not really the sort of place in which beavers are accustomed to live, and the idea was not simply to keep them alive but to keep them happy and working, too.

It was, of course, a tense moment when we put them into the terrarium. It seemed as though they would soon settle down. After a thorough exploration of their new territory they bathed in the concrete pool, and then went into their box to sleep.

After that I had a great deal to do. The beavers took no notice at all of the school bell that clanged in the corridor twice every hour. They slept peacefully until sometime during the last school period, and when the human pupils went home the beavers were in full activity, which continued without a break until toward morning.

We had wondered which of the youngsters were females and which were male. The two tame ones, Number One and Number Three, were very unlike in their ways. Erik had christened Number One Tuff, since it was a real "tough guy" and seemed to be the unquestioned boss of the whole group. Number Three was more retiring and not nearly so obstinate and self-willed as Tuff, so Erik thought a suitable name would be Tuss (a ball of cotton). The two other youngsters never talked to us and never ventured to take food out of our hands. In the terrarium they always kept in the background, and it often seemed as though they did not even notice that we were present. If we took them into an unaccustomed environment they immediately became as shy as quite wild beavers, and since we never had any more intimate contact with them we never bothered to give them names, but simply called them Number Four and Number Five.

It had become clear to me during the summer that it is not possible to determine the sex of beavers by the methods I had so far applied. I had considered the possibility of using X-rays, and the Army Dog Department's dog school gave me facilities for trying this method on the four young beavers. They were very interested at the dog clinic in these unusual patients, which neither bit nor scratched in spite of all that had to be done to them. The X-ray plates showed clearly that all but Tuff had penis bones. So she was the only female of the group, which was the exact reverse of what we had thought. It would seem that the female sex is the strong one even in the world of beavers.

The pupils at the school were, of course, very interested in the "beaver class" on the top floor. One could easily line up a whole class along the partition in front of the terrarium, and the beavers quickly accustomed themselves to these visits. At first even Tuff and Tuss were shy in front of strangers, but they soon discovered that there were likely to be apples and other tidbits when visitors came, and after that they stood up on their hind legs and begged inside the partition, or clambered over the pupils to see if they had anything on them. Number Four and Number Five waited around quite happily farther back inside the terrarium where they knew that no one could get at them. But if anyone, known or unknown, climbed over the partition into the terrarium complete panic arose among all the beavers, and they rushed headlong into their box.

They were most delightful when they sat in peace and quiet at the edge of their bath, washing themselves or eating their natural food. The washing procedure was the big number. So long as even one of the beavers was sitting straight up on his tail working over his fur with his hands, the spectators always stood silent, with their eyes glued to him (Figs. 22 and 23). The reason is no doubt that the movements seem so human.

The beavers' movements when they are eating bark or fine twigs, on the other hand, are something quite peculiar to them (Fig. 24). They are amusing because they make one think of some sort of machine. A piece of wood that is to be barked is grasped with both hands; one hand closes around it and the other grips it between the little finger and the other fingers of the hand (Fig. 25). The stick is revolved quickly and evenly with this second hand and at the same time moved slowly sideways as the gnawing teeth of the underjaw strip off the bark (Fig. 11). A pause is made at regular intervals for chewing with the molars, swallowing, and sharpening the teeth, and when the stick is almost barked the end of it is held against the palm of one hand so the teeth can work right up to the end. The stick is then thrown aside and another one picked up from the ground by one hand.

The finest twigs are eaten up completely. If they are very fine they are first bent double, and then held out obliquely with both hands and fed quickly into the mouth as the front teeth bite them with very quick movements into small chips, and when the mouth is full of chips a pause is made for chewing with the molars.

This quick biting with the front teeth makes a very loud noise, and when all four beavers were occupied in this way in the terrarium it sounded like a workshop with several machines going at once. Every now and then the incisors of the underjaw were sharpened against those of the upper jaw, and after that one could hear distinctly that the bite was better so that the "machine" could run faster. Beavers can keep their teeth in trim even if they never have an opportunity to gnaw. The idea that they often gnaw simply to keep their teeth from growing too long is quite fallacious.

When the leaves began to change color in the autumn the beavers went over entirely to a diet of bark, and their consumption of aspen branches was greater than I had ever imagined. It was easy to keep the terrarium clean because the excreta always landed in the water. It sank to the bottom and all there was to do in the morning was to empty the bath and scoop up the completely odorless cellulose mass. But it was heavy work cutting down aspens and carrying great armfuls of branches up all those stairs, and pailfuls of excreta down. The caretaker, Mr. Mohlén, offered to help me. At first he used to cut aspens in various places all over town, but soon we had to arrange for bigger-scale transports. The headmaster, who had a forest homestead, presented the wood and Mohlén and I used to drive in and cut a tractorload at regular intervals.

I used to lay the aspen branches in the pool. The beavers trimmed them down to short lengths and gathered a pile of them together on the platform before they started eating. They could remain underwater a very long time and could gnaw just as effectively under water as on dry land. If I wedged the branches firm under the water, the beavers would cut them free, showing that young beavers can get their own food from the winter store under the water outside the lodge even in their first winter. When a beaver gnaws under water his mouth is closed by two folds of skin, so that the water cannot get in.

The First Lodge

At the end of their first week in the terrarium the young beavers started carrying about pieces of barked wood and other material that was of no use for eating, in the same way they had done in Elias' workshop. If they saw a crack or a hole as they came along with their piece of wood they would try to stuff it in, steering it toward the hole with one hand and pushing it in, holding it between their teeth and knocking it in with energetic movements of the head, but they were not always successful by any means. The easiest place was the gap between the water pipes and the wall, and soon the beavers stuck all their barked pieces of wood there. They gradually started taking building material to their box, and when they had piled up enough sticks and branches there for it to be possible to fix sticks in the gaps between them, that form of building activity was also transferred to the area around their box (Fig. 26).

Normally they built only in the later part of the night and I had to sacrifice a good deal of sleep before I discovered what they did. Large branches were gripped between the teeth close to one end and dragged to the building site; smaller branches and sticks were lifted up to the mouth, one by one, by one hand until the mouth was full. Fine twigs, rubbish and earth were gathered together into a heap by simultaneous pushing movements with both hands, which were then pushed in under the heap so that it was held firm between the upper surfaces of the hands and the chin. With their hands pressed against the floor the beavers then pushed rapidly off along the platform with the heap in front of them. If there was any obstacle in the way they would raise their hands from the floor so that

29 Tuff at about two years of age.

30-32 When the lodge in the terrarium was more than three feet high and the youngsters had hollowed out their living chamber right up to the partition that divided the terrarium from the rest of the room, I made an observation hole in it. But when I opened the door to see what the youngsters were doing inside the chamber, Tuff always came up and nuzzled at me and then built up the opening in front of my face.

33 Tuff depositing castoreum on a tree trunk. She humps her back and kicks energetically with one hind leg, while audibly forcing out a stream of castoreum.

34 Two young beavers wrestling in the beaver dam.

35 The new-born baby is just dry. The female has not yet eaten up the placenta.

36 A few hours later the baby lay suckling beside its mother.

37 The female examines her baby with her nose while making her contact sounds. The babies start bathing on about the fifth day after birth, and after that they spend more and more of their time at the pool in the feeding chamber in the lodge.

38 Tuff always followed after her baby when it went out into the water to hold it by the tail while it bathed.

39 Mother and baby in their right milieu. Notice the long teat on the big, full breast under the female's left hand.

40 By the age of about one month the youngsters are getting quite independent, and the female then has more time to devote to the care of her own coat. In this photograph she is combing her breasts with the double claw. Their size indicates that the baby had not stopped suckling, even though he now handles sticks quite skillfully. The young stop suckling finally at the age of about two months. By then their movement patterns are in the main fully developed.

the burden was lifted elegantly over it, and they were able to raise or lower it slowly to the desired level, using their tails as balancers. If they had to clear a wide obstruction or go up a slope they would walk on their hind legs, carrying their burden between their outstretched arms and their chins (Fig. 27). They could transport large armfuls in this way and it looked very odd to see them lumbering along slowly and clumsily in an almost upright position with their toes turned sharply inward. Their fingers were stretched full out in front of the burden, their eyes seemed fixed on some point in the ceiling and their expressions were very purposeful and determined (Fig. 28).

The finer material was pressed firmly against the building structure with the flats of their hands or their chins. The building appeared to be planless but, to judge from the results, there must have been a certain system in it all, and I began to suspect that the animals unloaded and fastened on material wherever they saw a hollow or crack anywhere near their box.

Before going in to sleep in the morning they carefully cleared away all refuse from the platform, and when the building was most intensive a good many unbarked pieces of wood got carried down to the building site.

The beavers built only when they felt themselves completely undisturbed, and only because of the observation room was I able to watch and film the transportation and building work. Tuff was always very friendly and talked to me a good deal when I was in the terrarium room, but build while I was there she would not.

At times all the beavers were very shy, and at these times I could hear them all rushing into their box in panic as soon as I opened the door of the long corridor at the far end of which lay the terrarium. But if I went into the observation room and sat there quite quietly, they would soon come out again and resume their activities. In the evenings I often had visits from people who wanted to see the beavers, and when they were in their sensitive mood I would see them suddenly sense danger and look uneasy as I sat in the observation room.

In the end they would suddenly rush into their box, and a little while later some visitor would knock on the door. It seemed that the beavers could hear when the outer door was opened four stories down, which indicates that they have a capacity for detecting sounds which is so superior to that of humans that it is difficult for us to imagine what the hearing sense can mean to them.

The building activity increased night by night until the pile of building material occupied the whole space around and above the box right up to the platform, which was twenty-eight inches above the floor. Only a narrow passage was left along the wall to the entrance to the box. Then the beavers started hollowing out a cavity in the center of the compact pile.

One morning in the beginning of November, Tuff moved over from the box to the new cavity and a few days later the males, too, started using it as a sleeping place. It was then quickly expanded until it was more than six feet in diameter and sixteen inches high.

All this time the beavers went on building onto the pile from outside. If they saw no crack or hole when they came along carrying their building material they would unload it onto the middle of the pile, which consequently assumed more and more the appearance of a proper beaver lodge. On the side that faced onto the water tank, and from which the animals had to come carrying their building material, all cracks and uneven places were filled in with finer material, but the back wall of the lodge consisted simply of woven branches and pieces of wood that let the air through. The lodge was soon more than three feet high and more than six in diameter and it contained a great quantity of wood.

The young beavers, which were still nowhere near half grown and had never seen adult beavers build, had built a proper beaver lodge on the floor of a room.

When the chamber inside the lodge reached the partition in the terrarium I made a hole in it, so that I should be able to watch the beavers even when they were inside their chamber (Fig. 30). But when I lay on my stomach on the floor and peeped in, Tuff always came up and nuzzled at me, and then went away and fetched an armful of building material with which she carefully stopped up the opening in front of my nose (Figs. 31 and 32). Out in the terrarium she was as friendly as could be, but peering into her sleeping chamber she would not tolerate. I outwitted her in the end by putting a pane of glass in the hole.

The Territory

Even after the beavers had settled down in the terrarium, Tuff continued to inspect her territory, and on these regular tours of inspection she would nose around the lodge, on the walls between it and the tank and over a few tree trunks that lay on the platform. One evening toward the end of October she started sprinkling castoreum on the tree trunks; she placed herself astraddle them, hunched her back and kicked vigorously with one hind leg, audibly forcing out a stream of castoreum (Fig. 33). The males did not inspect so regularly. If they happened to pass a place that Tuff had just sprinkled with castoreum they would sometimes add a little too, but often they could merely draw their out-turned "cloaca" over the tree trunks as they passed.

At the end of October a young female beaver was caught on Faxälven. Since it was much too late in the year to let a beaver loose on strange ground, she had to be housed in a small room at the school. I was not sure that we would be able to keep her alive over the winter. We had found that it was very difficult to keep wild beavers alone in captivity for any length of time. All beavers, with the exception of very young ones, are completely apathetic for the first few days in captivity, and it is difficult for a solitary one to come out of that condition. One-year-olds in particular, if they do not get companionship, may simply sit where they are put down until they die.

This young, wild female was a very beautiful specimen, and Mohlén, the caretaker, christened her Fina. At the end of a month she made contact sounds as soon as I came into the room and after that she quickly adapted herself and grew plump and smooth-coated. It was obvious that she appreciated my company, but I was never allowed to touch her and it was only with the greatest hesitation that she ventured to take apples from my hand.

Tuff was often allowed loose in the Biological Institute premises while I was busy there. She always got very excited whenever she came anywhere near Fina's room. If I let her in she would try to get at Fina. And there was no doubt at all about her intentions. She flicked her tail along the ground in an irritated way, sharpened her teeth so that one could hear them grating against each other and had such an aggressive look on her face that one hardly knew her. Fina, on the other hand, gave her contact sound. She was always very interested as soon as she caught the smell of other beavers, and indicated clearly that she was always ready to seek contact with strangers.

When I prevented Tuff from attacking Fina she went for me instead, and if I then raised my foot at her she gripped it with both hands, put her chin against it and pushed with all her might. As long as I exerted myself and offered resistance Tuff would go on fighting, but if I gave in she would quieten down at once. If I did not let her work off her aggressiveness before lifting her back into the terrarium she would set on one of the males and wrestle with him until he gave in (Fig. 34). A wrestling match like this could go on intensively for ten minutes, and often ended with Tuff pushing her opponent down into the water.

Tuff's wrestling matches with me or with the other members of the group always ended amicably, as soon as Tuff had been acknowledged victor.

To see how the young beavers would react to strange intruders

I sometimes put beavers of a different sex and age in a cage, which I stood in the terrarium. The males did not react at all to the strangers, but Tuff always behaved extremely threateningly and never let them out of her sight. In the end she would get hungry, and then she would hurry off to fetch food, which she brought back to her place beside the cage, so that she never need let the intruder out of her sight while she ate, and every now and then she would stop to flay her tail and gnash her teeth at the stranger, who promptly lay flat on his stomach and looked unhappy. As Tuff could not get into the cage to settle matters with the intruder she had in the end to work off her aggressiveness on another member of the group, who always had quite a bad time on such occasions.

One evening when Tuff was loose in the Biology Institute she suddenly started sprinkling castoreum all along the walls of the corridor. I knew by that time what this meant and was a little uneasy. Quite right. When I lifted her back into the terrarium she stood up on her hind legs, looking out over the partition, and remained like that all the time I was in the room. The following morning I saw that my fears had been justified. During the night Tuff had gnawed through the partition from the top right down to the bottom. She had then gone out into the room and started gnawing a hole through the door out into the corridor, which she had obviously incorporated in her territory the evening before when she staked her claim by marking it with castoreum. But, mercifully, morning had come before she got through the door. As she could not climb back onto the platform, she had gnawed an opening from the room directly into the lodge, where she was lying asleep, pressed tight against her group companions when I arrived. I nailed a metal sheet over the hole, and when Tuff came out in the evening she went straight to the hole in the partition and tried to tear the sheet away with her teeth. When this failed she resigned herself, and fortunately she contented herself after that with her old territory.

Digging Without Earth

The beavers had no facilities in the terrarium for digging in a normal way, but in spite of it they dug at regular intervals. Quite suddenly one of them would drop whatever he had on hand and march off purposefully to a particular corner where he would dig against the tiled wall for five to ten minutes. Never did the beavers look so happy as when they were standing on two legs in the corner, digging away. They only used two of the movement patterns that make up the beaver's normal digging procedure: they scratched, and every now and then they made a pushing movement with their hands. After digging for a while they would return to whatever they had been doing before, just as suddenly as they had left it. They would dig like this particularly often while the lodge-building was at its height. It seemed as though they had a need to work off a digging instinct, and that was the strongest at those times when beavers in the open were digging particularly hard.

One evening I set up a wooden board to shut off the terrarium in front of the lodge. It was not long before the beavers wanted to get in, and when they found the way barred they started digging in the corner between the board and the Masonite floor. Tuff was particularly energetic. She would scratch first for a long time with her hands, and then make a series of alternate shoveling motions with her feet, which were more effective because of the webbing. After she had alternately scratched and shoveled for a while, she would make a sweeping movement to the side with one arm and finally turn completely around, stretch her arms forward and push the loosened earth—which did not exist—toward the wall, where she packed it tight with her hands. Then she hurried back to go on digging.

After a time she got the idea that she could tear up the Masonite on the floor with her teeth. To stop this I laid down a metal sheet, and after that the digging went even better;

while shoveling she rested on her hands and swept her tail at great speed over the smooth sheeting between her back legs.

The males contented themselves on the whole with scratching and shoveling, but every now and they went backward, scratching intensively, and it would no doubt have been an excellent method of getting rid of the loosened earth—if there had been any.

The young beavers went on digging by the hour so long as the board remained in position. It was only when I offered them an apple that they allowed themselves a pause.

The whole thing was almost laughable. They seemed to be functioning like some sort of elaborate digging machines. As soon as they met the obstacle something obviously pressed the button, and after that the machines went on working until I switched them off by removing the obstacle.

I had never at this stage seen beavers digging passages through the ground, but I could very well imagine that they did so in the way in which these young ones had tried to get through the board. The movements seemed particularly suitable for digging in narrow passages. I knew from my reading that similar digging movements occur in other digging mammals, but I had never read that all the movement patterns could occur in one and the same species of animal. Beavers were obviously particularly skillful diggers.

These beavers had never had the opportunity to dig passages, and yet the whole chain of digging motions proceeded in correct succession as soon as the beavers were presented with a sufficiently strong urge to dig.

The reason Tuff dug more intensively than the males and exhibited a more complete series of digging motions was presumably that she was more closely bound to the territory and to the lodge, so that her need to break through to it was particularly strong. She had already shown in various ways that the territory meant more to her than it did to the males.

The winter passed quickly. The beavers grew and soon each weighed about twenty-two pounds. It was no longer so easy to have Tuff on one's knee.

Fina and Esmeralda

In April the first beaver of the year was caught on Faxälven. It was a splendid elderly female and weighed fifty-five pounds. As there was still ice and snow on the beaver grounds in the woodlands I wanted to try to keep her until spring was a little farther advanced. Now Fina could have the companion she seemed to have been longing for for so long, and afterward both animals could be let loose together.

When we took the big female into Fina's room in the morning, Fina was asleep in her box. I moved her over to a hollow she had gnawed in a pile of branches on the floor and closed the entrance so that she would not be able to get out, and then I popped the strange female into Fina's box.

I went back in the evening and let Fina out. She, not at all suspicious, went to the swimming pool and bathed. When she had finished her evening toilet she sat down calmly to feed, but soon her attention was drawn in the direction of her box and in the end she went over to it. She went in without hesitation. A slight scuffle could be heard going on inside and soon Fina came out and went back to her feeding place, but she seemed a little uneasy and not to have quite the same appetite as before. Every now and then she would stop eating to look over at her box. Before long she went over again, but stopped at the entrance and sniffed cautiously at the strange beaver, after which she trudged slowly back to her feeding place beside the water.

Several hours passed like this, with Fina going over at regular intervals to peep shyly and cautiously into her box, but at last she got her courage and went in. Immediately there was a terrible commotion. It seemed almost as though the box was going to explode, and out came Fina as though she had been shot from a gun. She lay down in a corner, pressed her chin on the floor and looked very dejected. I felt quite sorry for her. The stranger whose acquaintance she would so much have liked to make had hit her.

As nothing happened for a long time I went home to have some coffee. When I came back an hour or so later Fina was sitting by the swimming pool eating as though nothing had happened. She seemed very self-possessed in comparison with earlier in the evening, and when I went into the room to find out what had happened while I was away I was met by a remarkable sight. The entrance to the box had been carefully blocked up from the outside with shredded wood and twigs, and outside that was a well-built fence of woven branches. Fina had shut the stranger into the box, and she seemed to have got it into her head that it was a truly bad beaver she had to deal with, for the outside of this well-constructed barricade was made up of the thickest sticks available in the room. Every now and then she stopped eating to do down and fix yet another branch into the bastion.

I would have liked to have stayed with the beavers all night, but in the end I went home. Next morning, with my heart in my mouth, I opened the door to the animal room. I was afraid it would look like a bloody battlefield inside, but everything seemed in order. The barricade in front of the entrance to the box was gone, and when I opened the lid I saw both the beavers lying there together. Fina was looking very calm and content and she seemed to be positively purring, lying there pressed close against the big beaver. They had obviously made friends during the night and I would have given a great deal to know how it happened.

After that Fina never stirred from the big beaver's side. It was days before the big one ventured out of the box, and all that time Fina stayed with her. She came out only now and them to bathe and get food for both of them.

Mohlén christened the newcomer Esmeralda. Beavers ought to have dignified, old-fashioned names, he thought, and since this particular beaver was such a typical stately matron, I could only agree that the name was suitable. Esmeralda soon settled down, and later on she and Fina became pioneers in a beaver ground that had once long ago belonged to their ancestors.

When I tell the story of how Fina shut Esmeralda into the box, people are inclined to think it shows that beavers are very intelligent. Actually Fina acted with pure instinct. Holes that offer certain stimuli quite simply actuate the beavers' building urge. When Tuff stopped up the observation hole in the lodge wall she did so because she could see a hole from which came an unpleasant smell, and I was able to outwit her by covering the hole with glass because then she was no longer aware of the unpleasant smell.

Joys and Disappointments

The scholastic year had given me more than I had even dared to dream of. The beavers had thrived and worked and remained typical beavers, in spite of their abnormal "upbringing" and in spite of the unnatural conditions. I had been able to make personal acquaintance with beavers, and I had got an insight into many aspects of their behavior that had been unknown. As always the number of question marks increased with every new detail learned, but the beavers' two terms at school had left me with high hopes for our future collaboration.

In the course of the autumn the hopes that fish and beavers might be able to return to Holmeselet had been dashed. The provisions about building an impounding reservoir had been the only salve for the wounds inflicted by the building of the power station and the river regulation, but the petitioners had put forward fresh demands as work on the power station proceeded.

In spite of energetic and unanimous protests by the villagers, the provisions for an impounding reservoir were rescinded. The applicants were given permission instead to make a timber-dumping ground on Holmeselet. This meant that the river bed would be full of water during the summer. Tourists would see no blot upon its beauty to complain of then, but in the autumn the water level would be dropped drastically, and when it was raised again in the spring the timber that had been laid up on the exposed bed during the winter would float off into the water without the use of labor. It was all perfectly rational, and no one thought about the villagers who alone had to pay dearly for the profits of rationalization from which others would benefit. Many Ramsele men would be out of work, and those who were able to stay on in their native village would have to watch their lovely bit of river converted every autumn into an ugly stony desert. The once excellent fishing would be ruined forever, and the beavers that established themselves by the river in the summer would lose all possibility of existence in the autumn.

In Pastures Green

All that winter I had been searching for a suitable section of stream that I could fence off for the beavers, where they would have an opportunity to show what they could do in a natural setting. It was more difficult to find a really suitable place than I had expected, but in the end I heard of a homestead for sale a few miles out of town. We drove out to look at it and found the most ideal situation for beaver dams it was possible to imagine. The homestead was beautifully situated on a slope facing the Ångerman River. We had nothing against living there and the place was ideal for beaver experiments.

We moved out in the spring, and as soon as we had got our belongings more or less in place we started making arrangements for the beavers.

We built an enclosure eighty feet long and about as wide around one of the streams close to its effluence into the river. The big barn stood empty and there we fitted out eight stalls, each equipped with an old discarded trough to be used as swimming pool so that we would be able to receive the newly caught Faxälven beavers. We intended to start the trapping in the summer.

A Stockholm schoolboy had written during the term to ask if he could come and help us with the beaver experiments during the summer holidays. Tryggve arrived as soon as school was over, and his first task was to get acquainted with the young beavers at the school. They had stopped building long ago and were spending most of their time growing fat on the first fruits of spring. Buds and freshly opened leaves were obviously a real delicacy after the long winter, with its diet of bark.

The very first time the beavers were offered leaves that year, they handled them completely differently. They gathered them together into a bunch between their hands and then fed it into their mouths with the leaves still on the twigs. This was much quicker than the way they had delicately bitten off one leaf at a time in their first summer and autumn and rolled them neatly into a cigar shape. The new movement pattern had obviously matured in them during the winter, when they had no access to fresh food, and was ready to be applied as soon as the trees broke into leaf in the second year of their lives.

Early one June morning Tryggve and I carried the beavers down to the enclosure. We had brought them from the school the evening before, and they had had to sit all night in a cage so small they had not been able to groom their coats properly.

They nosed around on the ground in the enclosure for a time, but soon found the water. For Tuff and Tuss it was the very first time they had ever been able to move quite freely in such a large space out in the open.

The many new stimuli started an intensive activity in all four beavers; they swam restlessly up and down in the stream and ran up and down the slopes of the bank. After a while Tuff sprinkled castoreum on piles of earth on the shore close to the net where it closed off the stream, both where the stream ran into and out of the enclosure. She was obviously already so much at home in the new territory that she felt it time to mark it with her stamp of ownership.

As the beavers had not been able to groom themselves the night before, they got wet to the skin and looked miserable scampering around on the banks. It was some while before they quieted down enough to wash themselves clean, but after that they gave themselves a special grooming so that toward afternoon they were looking more like beavers again.

After that they started digging with greater energy than ever. Whenever they passed a hollow in the bank as they swam along, they would steer over to it to see if it was a suitable place in which to dig. They were so eager that they sometimes started fighting over the best working places. If two of them caught sight of the same spot, the first one there made threatening noises until the other one went away to find another digging place. Before long there were passages started in every hollow and recess along both banks.

Some of the passages opened out under the water, and after a time they dug only in these. It was particularly exciting to see the way the beavers dug in the earth. Unfortunately I could not see what they were doing inside the passages, but when they started digging a new passage they would scratch and shovel alternately, and after the passage had deepened so that I could no longer see the digger he would come along to the mouth at intervals pushing a large pile of earth in front of him with his hands. When all the animals were busy, each digging a different passage, one cloud of earth after another kept bursting out of the mouths of the passages, until the previously clear stream became quite muddy. It seemed that the animals were applying exactly the same method when they dug passages as when they dug on the floor in the terrarium in front of the obstacle that barred the way to their lodge.

They were active day and night for the first forty-eight hours and never even allowed themselves to sit still and eat for any length of time. On the second night they investigated their land area, and once they had established its limitations, took no further interest in the net. On the fourth morning they installed themselves in a cavity inside the longest passage, which led down to the bottom of the stream.

Tryggve was to watch the beavers for the next few nights. In the evening he set up a tent inside the enclosure, but when he went in to have a little sleep, it was not very long before Tuff came in to see what it was that had suddenly grown up inside her territory. Tryggve fell asleep after pushing Tuff out and carefully closing the tent opening, but before very long he was awakened again by a beaver tumbling down on him from above. It was Tuff, who had climbed up onto the tent and brought it all down with her weight, and it was so badly damaged that we could not use it again. Tryggve got no sleep at all that night.

It was a week before the beavers seemed really at home in their enclosure, which by then began to look like a real beaver haunt. The plant life had been severely eaten and here and there young trees had been felled. Clearly defined roadways led down to the water, where the many feeding places at the water's edge were well worn.

Once the beavers were installed they grew more timid and no longer allowed one to stand inside the enclosure to watch them. They soon, however, became accustomed to people standing outside the net. If one went inside they behaved like wild beavers, striking their tails against the water and diving into their passages.

One evening only Number Four and Number Five were about when I went down to the enclosure, and for the next few nights we still saw nothing at all of Tuff and Tuss. We began to get really uneasy and in the end we captured Numbers Four and Five in order to make it easier to tell whether Tuff and Tuss were still there. After that the enclosure seemed completely lifeless. We were about to begin a desperate search of the surrounding country and were in complete despair. We were convinced that the animals on which we had built such hopes were gone forever.

The only remaining possibility was that they might have hidden themselves somewhere inside the enclosure, and Tryggve and I decided to sit and watch there the entire night. Gloom descended over the beaver stream that night and we had abandoned hope when at about three o'clock we saw Tuff cautiously creeping out from a passage. She collected a little food in the stream and then went back just as cautiously as she had come. Of Tuss there was no sign at all.

We dug Tuff out and we soon found Tuss, too, in the same passage. Number Four and Number Five had stayed in the second of the two longer passages, but to start with all the animals had slept together. Presumably Tuff and Tuss had later retired to their own passage, where they had isolated themselves from the two "superfluous" males.

We put Number Four and Number Five back in the enclosure, but Tuff and Tuss had to go to the Hölle laboratory on Indalsälven, about thirty-five miles from home, where they were to live in a large stream aquarium. There I was able to film them through a glass wall as they swam under water, and I had a faint hope that they might build a dam in the aquarium.

Beaver-catching went on at full speed that summer. At times there were sixteen beavers munching aspen branches in the barn at the same time. Because we caught so many we were able, with the help of X-ray, to pair the animals off quickly. If a male and a female were put together in the same stall simultaneously they soon accepted each other. After that it was vital to let them loose as soon as they had got to know one another well enough to remain together after being liberated. If they were kept in captivity too long, they lost weight and their fur grew dull and lusterless.

There were one or two tragedies in the barn. One young male had to be left without a companion and after a few days he was found dead in his stall. One pair broke in to another pair and the animals fought so fiercely that all four were badly hurt. Two of them we managed to save, after the vet had sewn them up, but the other two died. Still, thirty-four Faxälven beavers were transferred safe and sound to new ground that summer.

In early July we dug out two more young beavers from different wood lodges, and later in the month another one. The first two were about as big as Tuff and Tuss when they were first caught. We X-rayed them and decided that they were all females. Tryggve christened them Eva, Greta and Stina. We put them all in an enclosure at home and they were just as delightful as Tuff and Tuss.

From the start Eva and Greta got excited, sharpened their teeth, switched their tails and behaved as though they were depositing castoreum as soon as they caught the smell of adult beavers. But although they pressed out the "cloaca," hunched their backs and kicked with one hind leg, no castoreum came. Their pouches were not nearly developed enough and were quite empty.

However excited the beavers got, one could never make them bite. Beavers have a strong social inhibition which prevents biting, and even if one puts one's fingers into a beaver's mouth when it is eating it will skillfully avoid hurting them. Only after they have been taking aggressive action for some time against strangers will they use their teeth as weapons, and then they normally direct the blows against the rump.

We built yet another enclosure in the course of the summer. In this we placed a captured pair of adult beavers, but of course they were so timid that it was impossible to see what they were doing at night.

During early August exciting things began to happen in the two males' enclosure. I discovered that they were fixing sticks every night where the water rushed through the netting at the inlet of the stream into the enclosure. Soon the water was gushing harder than ever through a wattle of woven branches, and then the animals tried to plaster it over with earth and other finer material. As they were building against the current it was not easy for them to get the material to hold. They did, however, succeed by degrees in raising the level of water a few inches, but of course on the wrong side of the netting.

At the other end of the enclosure where the water ran out, it again dashed through the meshes of the netting, and when the sound of running water at the inlet of the stream had been deadened by the damming there, the beavers went over to build in the same way at the outlet. There their labors met with better success, as they were able to work with the current.

The water rose rapidly and soon the beavers received unexpected help. It started raining, and went on raining continuously for a week. The little stream overflowed its banks so that the whole enclosure lay under water with the exception of the particular corner in which the beavers, "cleverly" enough, had built their living cavity at a safe distance above the water.

They seemed very pleased with the results of their dam-building, splashing happily away in the water, and soon they started collecting a big pile of branches on the ground, from the water's edge a little way up the shore, just above the spot where the passage to their living cavity ran deep under the surface of the ground.

How they could know that the passage ran just there is a mystery. Since the flooding the mouth of the passage lay deep below the surface of the water far out in the big dam, and at one point where they were building, the passage was about five feet underground. If they were being stimulated by the moisture inside the passage where it ran closest to the water's edge, they would then have to dive a long way through the part of the passage that was under water and rise to the surface, and then swim in to land in order to reach the spot where they were building. The fact remains that they built a proper river lodge and plastered it carefully with mud from the bottom of the stream. In their first winter they had built a stream lodge on the floor indoors, and now they were building a river lodge and reacting in exactly the same way as adult beavers when they settle on steep banks.

The beavers had just finished their lodge when the rain stopped and the water began to recede so that soon their lodge was lying high above the water in the middle of the sloping shore, where it was of no use. But they resumed work on the dam and before long had raised the water level about two feet above the level at which it had stood before the flooding.

I then laid a few pieces of thick rubber tubing to act as syphons over the dam; the water level fell and the dam lay exposed above the water. The two males then concentrated all their interest on the upstream side of the dam. They plastered with wet earth and worked away at that side of the dam until it was absolutely smooth, but after that they stopped, even though their work had not resulted in raising the water level.

They could not understand that the water was being carried across by the pipes.

Tuff and Tuss did the same thing in their artificial beaver stream in the stream aquarium. They first tried building a dam at the entrance to the aquarium channel, and then started building at the outlet, but this never resulted in any proper damming the whole time the animals remained in the aquarium. I thought it was due to lack of experience that the young animals started building in the wrong place, but it soon turned out that the old, experienced animals in the new enclosure behaved in just the same way. It appears that all beavers start building where the water flows fastest.

The rubber tubing over the young males' dam clogged up after a time so that the water began to rise again, and the beavers then carried up another layer of branches and earth onto the lodge, though it was still lying high up on dry land. When the water started running over the dam they built that up higher, until the flow of water over the crown had stopped. The young beavers obviously "knew" that the overspill ought to be at the sides of a beaver dam.

The two males had felled young trees all through the summer, but only after the dam was completed did they start felling wood in earnest. They built a winter store in the water in front of the entrance to the passage, and before long they had collected such a pile of branches that it almost blocked the stream just downstream of the passage entrance. They had fastened finer twigs between the branches so that the pile was very dense. Only a few twigs appeared here and there above the surface of the water, but after I had cleaned out the lengths of rubber tubing so that they began to suck water over the dam again a large part of the winter store became visible above the water that started rippling between the branches. This was a stimulus the beavers could not resist and they started carrying barked pieces of wood and other building material to the store. After they had plastered it with earth on the upstream side they had converted it into a dam that raised the water several inches higher outside the lodge. It was a shame to waste that fine winter store, which would certainly have been enough for both animals all through the winter.

When I saw the males converting their store into a dam, it struck me that this must be what the Faxälven beavers do after they fix alder branches on top of their store. When the store reaches up to the surface of the water, the fast-flowing water, of course, eddies around the branches that stick up above it, and presumably that is the stimulus that makes the beavers switch over to collecting building material.

The autumn had begun and I thought it was safest not to disturb the beavers in their autumn occupations. So I removed the rubber tubing so that the winter store again stood above the surface of the water. Fortunately the beavers immediately started tearing away all the building material from the store, and after that they dragged all the material that was of no use as food to the dam on the outlet side, and collected all the eatable wood in the store, which had grown very large by the time that ice began to form on the beaver dam.

The enclosure had at first been thick with young trees, but by the time autumn came it had been pretty well cleared, so that we transported branches there which the beavers immediately dragged to their store. To find out whether they were able to fell larger trees, we set up a birch tree, about four inches in diameter, in a vertical position. The beavers felled it the following night, and they did their work so beautifully that even an old, experienced beaver would have had nothing to be ashamed of. The young beavers had evidently mastered the art, even though they had never attempted a thick tree previously, and even though the work called for a quite different technique from the one they had applied in cutting smaller trees.

Tuff and Tuss were making no great progress at Hölle, and at the end of September I brought them home. As soon as I had finally got hold of them in the aquarium one could see that they still recognized me, and Tuff readily sat on my lap and ate an apple. She weighed just under thirty pounds.

The family was pleased to see her again and we kept her loose in the house for a time before putting her and Tuss back in the enclosure. We caught the wild beavers that had been living there and turned them loose in a wooded stream not very far from home.

As soon as we put Tuff and Tuss back in the enclosure they behaved like wild beavers and never showed themselves. I wanted them to build a proper beaver dam, not supported by any netting, so I piled up stones and branches in the middle of the stream until the water splashed more there than it did at the inlet and outlet of the stream. My calculations proved correct. That very first night in the enclosure Tuff and Tuss went on with my work and by morning they had achieved a perfect, absolutely watertight, beaver dam. It was good to be reassured that they, too, knew their stuff, but it was disappointing to have no further contact with our old protégés who had once seemed so affectionate.

Winter: Under Ice and Under Roof

Tuff and Tuss were overtaken by the winter before they had had time to build a lodge and gather a store, and they would have found themselves in a very precarious situation if they had been entirely dependent on themselves. Fortunately they had had time to dig burrows that reached up almost to the surface of the ground, so they had ventilation in their lodge.

The beaver dams were covered with ice and the ground was hard. The frozen tracks from the young beavers' first summer out in the open emphasized the silence and emptiness of the enclosures. But underneath the ice and the frozen ground the beavers would be warm and snug through the winter.

Tuff and Tuss were dependent on our giving them food through the winter, but the males in the second enclosure could manage on their own large winter store. As we did not want to lose contact with them either we arranged for feeding in both enclosures. We placed a box with a double lid over a hole out in the ice, and through this we were able to push down aspen branches and hang turnips from hooks down into the water. The beavers were obviously swimming around a great deal in the water under the ice, for they found the food almost at once.

In December the beavers in both enclosures gnawed and dug a furrow in the crown of their dams so that the water level sank several inches, and after that it was easy for them to be out in the dam under the ice covering. It was a good sign, which gave us great hopes that they would survive the winter.

I had never known that beavers out in the open made openings in their dams before the late winter. It may possibly have been because of the severely restricted space that the animals in the enclosures opened their dams so early.

We would not see last year's beavers again until the spring, but I was far from being unoccupied through the winter. Eva, Greta and Stina were X-rayed again at the end of September and it was found that Stina was a male. We then put the female Greta and the male Stina into the Hölle aquarium. They were still quite small and I hoped that they would have time to adapt themselves to the narrow space sufficiently to build a dam in the aquarium the following autumn. It was not long before they started building over their living box and they soon converted it into a real beaver lodge.

With Greta and Stina gone to Hölle, Eva was left alone and we moved her down to the cellar of the house, where she, too, built a large beaver lodge on the floor. But Eva was not happy alone. If we forgot to close the door properly she would soon come clambering up the steps into the kitchen, and at times she got up to a good deal of mischief. For example, Ulla had a lot of hyacinth and tulip bulbs in the cellar, which were intended to be ready for bringing out at Christmas, but as soon as Eva found the pots she dug out the bulbs and ate them.

We were going away for Christmas. As Eva had stopped building and I did not expect to learn anything more from her that winter, we decided to take her back to her old companions at Hölle before we left. Greta appeared rather aggressive when we put Eva into the aquarium and the two animals fought for a time, but when Eva gave in and retired to the corner farthest from the lodge, Greta calmed down. But when Eva swam toward the lodge Greta attacked and this led to a violent struggle. Next morning Eva was lying dead beside the lodge with her spine bitten through just above the root of the tail. Greta had obviously dispatched her as she was trying to enter the lodge.

Of course we were very sorry about what had occurred, though at the same time I was grateful to have had such a clear demonstration of how inexorable the laws of the beavers are.

Greta and the male Stina had obviously established territorial claims over the aquarium that autumn. After that, Greta regarded all other beavers as intruders and if they came too close to the lodge no mercy was given. I alone was responsible for the tragedy as I had put Eva into a strange territory in which she was regarded as a mortal enemy. Such events are probably very rare in the open, if they occur at all. Beavers know the boundaries of their territory, and if they happened to enter a strange one they would be speedily reminded of the fact and withdraw, possibly without any serious consequences.

The pair of wild beavers that we had moved from the enclosure out into the wood wandered around for miles before, late in the autumn, they returned to the place where they had been released and settled down there just before winter came. They had not had time either to build a satisfactory system of passages or to gather a winter store. We cut a hole in the ice and stuck branches through into the water, but even so it was apparent after a time that they were in want. As soon as it was no colder than about twenty degrees, they would sit all day long on the edge of a hole in the ice eating, and they must have been very hungry for they were not too timid, and allowed one to come quite close to them.

At night they felled trees, and this they went on doing as long as they were able to get up on land. They seemed to be working in sheer desperation. In the end seventy-four trees two to eight inches in diameter and a few hundred smaller saplings lay felled but not trimmed within the area. As they were left there untouched, the labor seemed completely futile.

I had seen similar abnormal beaver fellings out in the wild on a number of occasions. They seem to occur in places where beavers have settled down so late in the autumn that they have not had time to build a lodge and collect a store before winter comes. Normally the lodge is built first, and it seems as though beavers cannot collect their store until that is finished. If they start the building work later, the collecting urge does not make itself felt until the lodge is finished, and then they start

41 The grooming ritual is generally begun by cleaning the nose. Then the whole head is worked over, the chest, arms, stomach, thighs and back in this order.

42 The baby prefers to eat from the same branch as its mother.

43 The male grooms himself while the youngster eats close beside him.

44 The youngster has bitten off a leaf from a twig and is holding it pressed between its hands while it eats. The older animals bunch several leaves together and eat them direct from the twig.

45 One of the "bachelors."

46-48 A beaver goes up to his companion and "dances." The latter is then immediately prepared to take part in social grooming. The skin is stretched between the hands while the under fur is carefully worked through with the teeth.

49 The foundation is laid for a dam.

50 Small trees are cut off in a few seconds and dragged direct to the water.

felling trees to provide material for the store. But if they are caught by surprise by ice forming before the lodge is finished, it seems as though they work off the whole collecting urge by felling trees. It is probably also a carry-over of the collecting urge that makes them sometimes fell trees in greater quantities than they need when they are first able to get up above the ice in the early spring.

I also had opportunities for studying beavers that winter at Skansen in Stockholm. A pair of wild adult beavers from Faxälven had been put into one of the dams there in the autumn. Even if I did not have the opportunity of going to call on them very often, I was able to establish that wild adult beavers build lodges in captivity in the same way as the wild young ones. The pair at Skansen also collected a winter store in the water, and both the male and the female marked the territory with castoreum.

Ångermanälven, too, is under short-term regulation, but as the variations in water level are relatively slight downstream from Sollefteå, we could not imagine that it would have any effect at all on our enclosures, which lay up small streams some way away from the main river. But during the winter a large dam was formed by the regulation ice and the water rose so that it penetrated into the two males' enclosure, where the water was high all day. When the water fell in the evening, it swept back around the sides of the beavers' dam and the ground was soon eroded away. In the end a great gaping hole was left between the enclosure and the river. Filling this up with earth during the winter was no easy task. We worked desperately, but we soon had to admit that, in the matter of building dams, we were not in the same class with the beavers. After a great deal of labor we managed in the end to repair the leak, but later, when the ice melted, a large slice of shore was washed away by the falling and rising water. When I got there one morning I saw an opening several yards wide under the netting out into the river and in it were distinct beaver tracks. The next few days I found beaver-felled trees along the riverside all the way up to Sollefteå, and I thought I should never see the two males again, but it soon turned out that they returned every morning to their lodge inside the enclosure to sleep. We repaired the damage so that in the future the beavers would have to remain inside the enclosure. They had shown that they felt at home there, and they were not in captivity in the real sense. Since then I have become accustomed to my beavers making expeditions into the open every now and then. It is very difficult to shut them in effectively, and if one of their passages falls outside the enclosure, they sometimes take themselves off on journeys of exploration into the surrounding country. But if they have come to feel at home in their enclosure one can hardly prevent their getting back in. It is only the young animals that at certain periods want to leave home.

In the spring a large wild beaver settled down on the river shore outside the enclosure and it tried several times to get inside the males' enclosure. It was presumably a female we had released earlier a few miles upstream.

In the Beavers' Nursery

In May the two males started going up on the shore again. They were almost fully grown now, but were obviously still good friends, slept together in the lodge and groomed one another as before. As the month went by they grew more and more restless. They wandered up and down inside the enclosure, and they dug frequently and tore at the wire netting in the corner closest to the river. As I had seen many examples of young beavers out in the wild leaving their home lodges and wandering far afield during the spring, I interpreted their restlessness as a sign that they had been seized by a strong urge to leave the enclosure.

The dam in which Tuff and Tuss lived appeared, on the other hand, lifeless during the light part of the day. The animals continued their underground existence all through the spring, and in the end we decided to try to recapture them to see how they were getting on. We had not seen them since letting them loose in the enclosure eight months previously.

On the morning of May 23 we started digging down to their passages. We located them bit by bit, and when evening came we had worked over a hundred yards of beaver passages but had not seen a glimpse of the inhabitants of the system. The whole enclosure was undermined with passages that branched out in every direction. In three places they passed under the netting and ran far outside the enclosure, ending blind just under the surface of the ground. The amount of earth the two animals had carried out into the stream during the winter and summer was considerable, in round figures more than four hundred cubic feet. All the entrances to the passages lay under the surface of the water, and where the passages rose to water level inside the shore they had hollowed out a feeding chamber and pool of water exactly as in beaver lodges out in the wild. The hollow in which the animals lived during the winter lay deep under the surface of the ground, above water level just at its highest point.

We couldn't break off our work until we had found the beavers, and late in the evening we got hold of Tuff in a passage that ended blind, but Tuss we didn't manage to catch until the following day.

We had expected to find a completely wild Tuff. She had been just as shy as a wild beaver after she and Tuss had isolated themselves from the two males more than ten months before, and we had not had any closer contact with her for almost a year. But when we took her home she wandered calmly around and nuzzled us all, and when I put the heavy lump on my knee and gave her an apple, she sat quietly eating it. She behaved as though she had never been away from us. It was obviously she who had done the major part of the digging in the enclosure for she was absolutely bald on her rump. Otherwise she seemed to be in fine form.

We put her in the terrarium at the school, where she had spent the greater part of her first year of life, but which she had not been in for almost a year. She obviously felt herself at home from the start; she first bathed in the pool, and then, after washing herself, went straight to the living box to sleep. We had long since pulled down the beaver lodge she had helped to build when she was small, but the box still stood in the corner. As she showed none of the restless, exploratory urge that is characteristic of all beavers that find themselves in strange surroundings, it was clear that she recognized her childhood home.

In the evening Tuff was sitting by the pool eating when I returned to the terrarium (Fig. 29). She came over to me at once and talked, and when she stood up on her hind legs I could see that she had four small teats on her breasts. Could it be possible that she was pregnant? To find out for certain we X-rayed her. In the car on the way to the vet's she managed to loosen the string tying up a sack of apples and before long she was sitting beside me on the front seat, eating. The X-ray showed clearly a large and fully formed fetus.

This was really a joyful surprise. Presumably she and Tuss had "got engaged" the previous summer, at the time they isolated themselves from the two superfluous males in the enclosure. The married couple had then lived a retired, underground life, and in the middle of February the animals had mated in the water under the ice covering. Now Tuff was clearly expecting her first-born almost any moment.

It was an exciting wait. I thought of keeping watch all the time with Tuff, but for one thing the behavior of the two males in their enclosure was beginning to be so interesting that I wanted to spend the light part of the night with them, and for another I did not know whether I dared disturb Tuff during the actual birth. Neither did I know whether it is customary among beavers for the male to be present during the period of confinement, and to be on the safe side we had put Tuss back into the enclosure after restoring it as far as possible to the state it had been in prior to our excavations. At first he wandered uneasily up and down all night. Once when he happened to find himself outside the netting, he immediately started digging desperately to get back into the enclosure. After I had lifted him in, he gradually reconciled himself to his grass-widowerhood.

In the terrarium Tuff was tamer than ever. As soon as I opened the door she came up to me. Her contact sounds were even more highly modulated than before, and it really sounded as though she were talking to me. She obviously appreciated my being with her and it was not only because I gave her tidbits. Even if she had all manner of delicacies inside the terrarium, she would always come over to me to talk. There was no question at all but that she recognized me and still regarded me as a member of the group, and it almost seemed as though I were serving as deputy for her better half while he was away.

One day after another passed without anything happening, but on the sixth day after her capture Tuff seemed a little uneasy when I peeped into her box during the morning. She was sitting up with her tail turned forward and her head bent down toward the "cloaca." An hour or two later I again cautiously lifted the lid, and there sat a little beaver beside Tuff! The baby was already dry and its eyes were open. It sat upright, just like an adult beaver. Beside it lay the placenta, moist red and shining (Fig. 35).

Tuff seemed quite calm. When I held an apple out to her she immediately started eating, and at that I cautiously closed the lid and hastened home to tell them about the happy event.

That evening the whole family came back with me to the terrarium. We could hear the little one whimpering inside the box. Tuff was still with it, but when I opened the lid she went calmly over to the pool to bathe and then sat down to eat. The little one stopped crying as soon as it knew it was alone. We picked it up to weigh it, but it was already so strong and lively that we had great difficulty in getting it to sit still on the scales. In the end we were able to establish that it weighed just twenty-five ounces. Apart from the fact that the tail was soft and downy and closely covered with fine gray hair, the baby looked just like a "real" beaver. We put it down outside the box, where it trotted restlessly around on the floor for a bit with fully developed movements.

Tuff soon went back to the box, where she first investigated with her nose the part of the floor on which her baby had just been walking, and then stuck her head into the opening and started talking audibly with the baby. The long series of contact sounds was particularly varied and had a quite new, motherly sound about it. After she had gone inside one could hear from the sounds she was making that she was fussing around the baby, but when I opened the lid to satisfy my curiosity she came out again and sat down once more to eat beside the pool. She was so completely calm that it seemed as though she was taking the opportunity to eat while she could leave the baby in my care.

Next morning the baby was lying suckling beside Tuff in the

box (Fig. 36). Tuff had been in with it all night and lay still even when I opened the lid. The placenta was gone. Tuff had obviously eaten it during the night.

In the afternoon she went out and bathed as soon as I came back, and after that she sat down to feed beside the water while I petted the baby. It hissed at me angrily and indicated clearly that it did not accept me as a foster father. When I moved it out of the box it searched up and down, but soon found the opening and went inside. After that it always went straight back into the box no matter where I placed it in the terrarium.

Later that evening, intending no harm, I went straight over to the baby without first greeting Tuff, and it was not long before she came over to see what was going on. When she saw me she became very uneasy, and although I promptly withdrew, she investigated the whole terrarium very thoroughly after first making sure the baby was still inside the box. It was a long time before she grew really calm again. I had obviously made a mistake in not making contact with her before going over to the baby. She was very sensitive to any disturbances around her home if she did not know who it was she could hear moving about there.

We thought it was rather cramped for Tuff and the baby in the box, and five days later we put a larger one on the platform close to the pool. We did not expect Tuff to appreciate the novelty, but when she came out in the evening she investigated it immediately, and then went over and fetched the baby and lay down to feed it in the new box. It was late in the evening before she came out again to eat, but then the baby started exploring its new surroundings and in the end found the opening and came marching over to the water. It went on even after it had passed the edge of the pool, plopped in and swam straight across. But as the pool was not full the baby had no possibility of getting out by itself, even though it grabbed with its hands over the edge and kicked desperately with its back legs. The whole thing had taken only a few seconds, but already Tuff was beside her baby in its peril. There was a grim, determined look in her eyes; she seized the baby's tail between her teeth and pulled it out into the water. It looked as though she was going to drown it, and I sat with my heart in my mouth. The baby was struggling forward the whole time, and each time Tuff took a fresh grip with her teeth it managed to progress a little way so that soon it was back at the edge, and just then Tuff took a fresh grip. But instead of pulling the baby out into the water she gave it a push with her nose so that it landed on dry ground. Tuff then returned to her feeding place, but she was not allowed to stay there very long in peace, for almost at once the baby plopped back into the water again.

After Tuff had helped it out twice more it settled down and started grooming itself. The first contact with the water had caused it for the first time to pay attention to its fur. It washed its head and combed itself down the sides just like a big beaver, but each time it tried to sit upright to comb its stomach it toppled over backward. In the end it gave up and started gnawing awkwardly at a piece of stick. Then Tuff picked up her branch and sat down to eat beside the baby. She talked to it frequently and every now and then went on a tour of inspection all around the terrarium. As long as the baby stayed beside the water she inspected the territory at regular intervals, and if she was not sitting close beside it she would go over to it about every ten minutes to talk to it for a little while. She was obviously keeping a check at every moment on where it was.

I was afraid that all was not as it should be. If the baby had been left in the box it was born in, perhaps it would have stayed inside until it was bigger. Tuff certainly seemed to be looking after it very well, but what would happen, for example, if it got stuck under the ramp that led up from the water to the platform? And it looked wet and miserable. Could it really be good for such a tiny beaver to bathe so much? To be on the safe side I moved the new box down onto the floor beside the old one and lifted the baby in, but it promptly went out again and scuttled straight up the ramp, onto the platform and over to the pool, where without a moment's hesitation it plopped into the water. Tuff did not react immediately, but before long she picked the baby up and groomed it long and intensively. She rolled it between her hands and systematically bit through its fur with her teeth, and after a time she carefully tried out a grip with her mouth right across the baby's body, dived into the pool and swam under water to the other side. The baby kept absolutely still and hung like a dripping wet rag from its mother's mouth when she climbed up out of the water.

I felt almost certain then that Tuff was going to kill her baby. I thought we had interfered with her too much and that that was why she was acting this way. Surely it could never be right for an almost newborn baby beaver to be treated so roughly. But there was nothing to do but sit with folded hands and hope and pray that Tuff would not harm it. She laid the baby down on the platform, and then after many unsuccessful attempts got the right grip on it again and dived once more to take it back to the feeding place, where it lay down at once to sleep. I was afraid it would be cold so I moved it into the box, but it went back immediately and plopped into the water again.

This time Tuff felt her way to a firm grip with her teeth on the baby's tail and dragged it out of the water and straight back into the box. The baby struggled and got stuck in the narrow opening, but Tuff tried time after time and in the end succeeded in pushing her way through with the whimpering baby. Once inside she settled herself comfortably and the baby immediately flung itself at a teat and started feeding ravenously. I had been on tenterhooks all the time the baby was in the water, but now everything seemed in order and I was able to breathe again.

The next night the baby stayed inside until nearly morning, when it suddenly appeared in the opening, and when Tuff started talking to it it replied and went up to the pool and out into the water. Tuff's eyes again took on their grim, determined expression, she swam up to the baby, grabbed it by the tail with her teeth and steered it resolutely toward the ramp; first she pulled it backward, and when, in its struggles to move forward, it happened to turn its nose toward the ramp, she pushed it in front of her and so got it at last onto dry land. Then she danced in front of it and it made its way back to the box. After that she inspected the territory, blocked up the opening in the box, bathed and finally tore away the building material and went in to the baby. It promptly started clambering over her for a teat, crying loudly when she rebuffed it, but she bedded carefully with shredded wood before settling herself comfortably on her back so that the baby could lie on her breast.

During the next twenty-four hours I came to recognize that Tuff was looking after her baby in the best possible manner. She talked to it all day long, while she was nursing it or while she was washing it long and carefully. She spent the evenings and nights by the water, but she went over to the box at regular intervals simply to feel the baby with her nose and talk to it (Fig. 37). She looked extremely contented as she sat beside it, giving out her long series of motherly contact sounds.

The baby came out to bathe for a time every night, and as soon as it entered the water Tuff hastened to follow it and hold it by the tail as it bathed (Fig. 38). After a while she pushed it in front of her up the ramp and as long as it was out she inspected the territory at regular intervals. It was difficult to understand the purpose of the tours of inspection inside the terrarium. Possibly she wanted to make sure that no intruders had entered the territory while the baby was outside the protected home, and her behavior would certainly have been reasonable if she had had a number of babies. She would then have discovered during her tours of inspection if any of her babies had wandered out of earshot. One baby she could keep track of so well that she always knew where it was, for even if I moved it to different places in the terrarium while she was not watching she always went straight to it when the need to seek contact with it made itself felt. After the baby had bathed a few times she picked it up in her mouth and carried it into the box.

Tuff was absolutely enchanting as a mother. She had a very expressive face and could look indescribably troubled and unhappy if, for example, some stranger came too close to her baby or if she heard unfamiliar sounds from her home. When she was watching over her baby in the water she could look grim and determined, but when she was nursing it in peace and quiet or fussing over it she radiated pure maternal happiness. While she was looking after her baby or when she put her head on one side and looked appealingly at me she could look really intelligent, but in spite of her wise look there were many things that indicated that she was acting from pure instinct, and she sometimes did things that seemed inexcusably stupid. If, for example, I held out an apple on one side of the pool and she was on the other, she would swim straight across to fetch the apple, but if the pool was empty she always climbed down into it and was then unable to climb up on the other side. She never learned to go around the pool.

Her abnormal surroundings meant that she sometimes found herself in situations in which her instinctive behavior was not able to function. If, for example, she was disturbed while her baby was out in the water, she would become frightened while at the same time she would feel a strong urge to protect her baby. This conflict situation sometimes led her to do the strangest things. For example, she would try to fasten pieces of wood onto the smooth floor, where there was no possibility of fixing them.

She did not become uneasy if I lifted the lid of her box while she was inside, but if I opened it while she was outside, so that there was no roof on her house when she came in, she would be very upset and fill the box with building material from outside, building so vigorously that she paid no regard at all to where the baby was. If it was inside the box it had to take care of itself as best it could while she pushed in long poles with all her might through the doorway. After bringing in all available building material she would gnaw out a hole underneath the branches, and then she had a roof over her head again. She would also build if I had a window open so that cold air was blowing into the room.

On the whole her instinctive maternal behavior functioned extremely well in spite of the abnormal conditions, and I was given a good demonstration of what happens inside lodges when beavers have young.

The first time the baby landed in the water it swam for a short time using all four legs, but after that it kept its hands still under its chin and moved its hind legs in the same way as do adult beavers. On the ninth night it tried to dive, but succeeded only in getting its head under water. Then it also started gnawing at an aspen leaf and did actually, after a long time, manage to swallow a little bit. A few days later it was swimming even under water with fully developed swimming motions and eating solid food with increasing skill.

It was soon so strong and lively that Tuff had difficulty in getting a grip on it when she wanted to carry it. It no longer appreciated being carried and would jump away as soon as she tried to fix her teeth on it. After that Tuff needed only to walk or swim behind it and toss her head to hurry it up onto land or into the box, but she still went into the water and followed after the baby as soon as she noticed that it was bathing.

By the time the baby was nineteen days old it was eating quite a large quantity of solid food. It liked to eat at the same twig or branch as its mother, but she preferred to be in peace while she ate (Figs. 42 and 44). She obviously "thought" that the baby was now big enough to get its own food. It spent a longer and longer part of the night by the water and acted more and more independently. Occasionally it would try to steal food from its mother and would "give voice" if she did not willingly hand over some tidbit. It ate the same sort of food as she from the start, including turnips and other things she had learned to eat after a long time in captivity.

One night the baby suddenly started carrying up branches out of the water and during the night piled the building material together into a heap on the platform. Now and then I saw it digging on the floor, and one day for the first time it pushed a stick into a crack in exactly the same way as an adult beaver. After that it was not long before it started holding sticks between its little finger and the other fingers when it ate bark.

By the time it was about a month old the female talked to it as to an adult beaver, that is to say the contact sounds no longer had the special motherly sound, and the baby could often be heard answering the female with the same sounds. When she groomed it with her teeth, it would groom its mother in the same way.

The baby grew more timid every night and started diving with a beat of its tail whenever I came into the room. The female no longer followed after it when it bathed, but if it dived with a beat of its tail she would go straight into the water to see where it had gone. She still kept her eye on it, talked to it often and let it suckle during the daytime, but it grew less and less interested in a milk diet.

The baby was quite playful, rolled around in the water and often danced by itself, sometimes trying to get its mother to take part in its games by biting her on the back or dancing in front of her, but adult beavers are not particularly playful. It took part more and more often in her tours of inspection around the terrarium.

But not until it was two months old could the youngster get along completely without the mother's direct care. It was then that it finally stopped suckling, and only then were the movements used in grooming, swimming and handling food fully developed. It started eating bark in increasing quantities and revolving the sticks between its hands as it stripped the bark. Its timidity was then fully developed and it could lie pressed to the bottom of the pool for fifteen minutes after I had come into the room before it would come up to the surface again and gradually resume its activities even though I was present. The female and the young one grunted at one another if they sat too close together while they were eating, and the relationship between them was in general as between adult, unpaired beavers in a group. After that the youngster would spend whole nights by the water. At the age of about two months it weighed seven pounds. It would certainly have been spending the whole night outside the lodge looking for food on its own account if it had been born in a natural environment, and by autumn it would have been ready to take part in the group's preparations for winter.

By the beginning of August we felt that Tuss had been a grass widower long enough, so we caught him again to reunite him with his spouse. I wondered rather how the meeting between father and child would go, for they had never seen each other before, and I was uneasy that Tuss might regard the youngster as a completely strange beaver.

For safety's sake I first put Tuss in a cage in the terrarium. Tuff immediately started threatening and grabbed hold of the cage and pushed it with all her might against the wall. The youngster got frightened and ran away and hid. Things looked far from well, but as I was anxious for the beaver family to be reunited before autumn, so that all the members of the family could take their part in the preparations for winter, I wanted to have a try at letting Tuss loose in the terrarium.

Tuff was sitting eating at the feeding place, but as soon as she heard the male walking about on the floor she grew very excited and soon went over to threaten and, in the end, attack him. The struggle became so fierce that I was frightened and tried to go between them, but then Tuff attacked me instead. She even used her teeth, but did not bite hard. Tuss took the opportunity to go away to the pool, but Tuff ran after him and tried to prevent his getting into the water. After that there was a wild hand-to-hand battle. The beavers were terrifically strong and I had very little chance of intervening. The youngster was afraid of its father, and if it came anywhere near its mother she swept it angrily away with her arm. In the end the poor little thing looked absolutely terrified.

Tuss gradually retired, and when he looked really chastened and withdrew meekly into a corner, Tuff calmed down. After eating for a little while and sprinkling castoreum on a tree trunk she went over to him, and when he timidly gave way she started grooming him and he promptly groomed her. The two animals spent a long time grooming each other, looking more and more peaceful and contended. Then they went and had a swim, but Tuss was tired and soon went into the old living box to sleep. For the next few hours Tuff went over to him in the box once every fifteen minutes and the two beavers talked to each other audibly and contentedly for a little while, after which Tuff returned to her feeding place. The contact sounds had a tone I had never heard before. They were of a different character from the female's sounds while tending her young, but she sounded just as content when she was talking to the male. That is evidently the way married beavers talk to one another.

After it had been in the neighborhood of the box where the male had just been lying, the youngster prowled uneasily up and down and behaved for the first time in its life as though it were depositing castoreum. As it had not dared to eat since Tuss came, it was very hungry, but it was not until the evening of the next day that it ventured to the feeding place, where the male was sitting eating a branch. It approached cautiously, stopping every now and then to give a quiet, modest contact sound. The male took no notice at all, but when the youngster had come quite close he gave a good-humored grunt, and the youngster immediately hurried forward and started eating. From then on it often sat eating at the same turnip as the male. He showed extraordinary patience with the youngster, which attached itself thereafter more often to its father than its mother, as she was often a little irritated with the young one after she had stopped feeding it (Fig. 43).

The first few days in the terrarium the male slept by himself in the old box, but after that he moved into the new one with Tuff and the youngster, and all three lay pressed closely together throughout the day. After that Tuff did not take so much notice of me but attached herself to Tuss instead. Family harmony was then completely established in the terrarium.

Around the first of September we let Tuff, Tuss and the youngster out in the enclosure. Tuff immediately inspected her old territory and deposited castoreum around the boundaries.

Bachelor Life

While Tuff was bringing up her baby in the terrarium, the two males were leading a pleasant bachelor life in their enclosure. When the weather was bad I spent the whole night with Tuff and the youngster, but on fine nights I was generally back and forth between the school and the enclosure several times.

My observation post was on the slope above the beaver stream, which runs parallel with the river for a little way before discharging into it. Beneath me lay the enclosure and the murmuring stream and all the activities of the beavers, but my view was dominated by the river, which a little further on, flowed broad and powerful beneath a high wall of fir-clad banks. In the evening the water sparkled in the light. Timber glided past slowly in a never-ending stream. Mergansers swooped by, garrots whistled over the surface of the water and great flocks of gray gulls flew with heavy, rhythmic wing beats before the dark cliffs. The melodious cries of the gulls gave a wonderful quality to the valley, and I was often distracted from my beaver observations when the gulls started sporting and playing on the river; a whole flock of them would settle on logs of timber farther upstream, and there they would stand at attention, each on his log, as they came sailing by. After a while they would come flying back in a group to start all over again, and they would go on like this for a long time.

Now and then a salmon would leap in the river and remind me that not so very long ago it had been a river known for its salmon fishing. I saw two hauls, it is true, from my observation post, and occasionally through my field glasses I saw someone land a forty-five-pound salmon on the shore, but fishing was not really worth while. It was often simply a question of continuing until the fishermen had received compensation for the damage done by the river regulation.

There is an old grave mound up the slope a little way from the enclosure and I once asked Elias what he thought the river looked like when the old fellow who was buried there was alive. Quick as lightning he replied: "Lad, there were fish in the stream in those days!" Here "the good old days" are always associated with quantities of fish.

Minks and foxes ran around on the river shore and occasionally I saw an otter, but I looked in vain for any seals in the river. Farther downstream there had been a few during the summer, and in previous summers they had been seen as far up as Sollefteå. But they had not come up so far this summer and neither had I seen the solitary wild beaver whose tracks I had noticed now and then along the shore.

When I came up to Ångermanland in the summer of 1957 I was surprised that the nights were so light. Since then I have made use of the light summer nights for my beaver studies every year, and now I find it hard to imagine spending the summer months in more southerly latitudes. Besides, it is not only the summer that is better up here than in southern Sweden; the winter is decidedly so, and as soon as the light comes back in February the finest time of the year begins, culminating in midsummer. Then the nights are so light that one hardly notices that it grows any darker around midnight. It merely grows a little quieter for a time while the birds rest, but then the beavers are at their most active.

The two males came out of their lodge between seven and eight o'clock. First, rings formed on the water for a little while above the passage entrance. The animals were obviously sitting washing themselves at the pool in the feeding chamber. Then the surface of the water began to ripple and soon a beaver's head appeared. It kept quite still for a long time, with only the nose and eyes above the surface, but when the nose had reported that it was only I who, true to my habit, was sitting staring up on the bank, the back and tail came up too. A short swim or two up and down, and then the beaver clambered up on the shore. His companion followed soon after, having taken the same precautions, and soon they were both hunting about along the beaver paths for some suitable leafy twig. These were beginning to run short inside the enclosure and the beavers had rejected most of the young aspens and sallow bushes that still remained. However fresh and green a tree may look, there is no guarantee that beavers will approve of it. There were still plenty of plants, and sometimes the two males went and grazed like cows on the banks. The comparison is a poor one, for I have never seen cows or any other grass-eating animal go around so fastidiously picking out a bud here and a particular kind of leaf there. But if the beavers found a good patch of some plant that they regarded as a special delicacy, they would gather up a whole armful and sit down to eat it at one of their feeding places beside the water, to which they also dragged all the edible branches they could find. If it found something particularly delicious it was so anxious to keep it to itself that it would whimper all the time it was eating it, even if the other one was inside the lodge (Fig. 45). One of them kept diving indoors to sit and eat in the feeding chamber; a lot of barked sticks collected there in the course of the night.

The beavers were restlessly active all night. Now and then they would seek each other out to get help with grooming their fur, and after that they would gnash their teeth together just like horses when they are similarly occupied (Figs. 46, 47 and 48).

The beavers were at their most peckish around midnight, but all of a sudden they would stop eating to march off quickly and determinedly to the corner nearest the river, where they would dig and shake the wire netting for a while. When they had had their fill of that they would return just as purposefully to the feeding place. Sometimes they passed within a foot or two of me without giving me a glance but hissing so angrily that I would have been afraid if I had not known that I could have put my fingers in their mouths without any danger.

Toward morning they spent more and more time playing in the water, and sometimes they gave showy displays. One of them would dance in front of the other, which would then fling itself into the water, pursued by the one that had started the game. They would go on playing clumsily under water, and then, swimming very quickly, spin rapidly on their lengthwise axis. It was all extremely graceful and showed how extraordinarily skillful beavers are under water. They are just as quick and supple there as they are heavy and clumsy on land. After playing for a time alone or together they would dive into the lodge, but before settling down inside for the day they would carefully clean out the feeding chamber. They carried out armful after armful of barked pieces of wood, which they dropped into the water a little way upstream from the entrance to the passage, until the floor of the stream was absolutely littered with wood. Some of it was carried along by the current for some distance before sinking in shallow water downstream from the lodge, where in the course of the summer quite a pile of barked wood and branches collected across the stream. Little by little the beavers took to carrying branches and sticks there that were of no use for eating, and in this way a solid base was soon formed for a beaver dam, though the actual building of the dam was not started during the summer (Fig. 49).

Since there were so few deciduous trees in the enclosure, I took some young aspens there and stood them up firmly in a natural position in front of my observation post. The young beavers had shown quite early in their first autumn that they were capable of felling the small trees that I had set up in a vertical position in the terrarium, but they were not particularly keen on that sort of work. They preferred to have their food served on the ground, and obviously thought it was more difficult to cut up trunks that stood upright than trunks that lay horizontally. They applied the same technique in both cases, except that they had to put their heads sideways in a particular way when they were gnawing through vertical trunks. Unless compelled to, older beavers also do not fell trees in captivity.

The two-year-old bachelors, on the other hand, had nothing against forestry work, and when they came out in the evenings they discovered at once that new trees had grown up during the day. They felled them before the night was out, but, the trees were standing up again next day. If the animals had had the capacity to think, they would have been particularly surprised to see that the trees grew thicker and thicker every time. Beavers do not normally fell large trees during the summer, but by gradually increasing the diameter I was able to draw the two males on, giving me an opportunity to study their working technique with trees of all diameters up to ten inches while the nights were light.

Young trees that were not too thick for the beavers to cut through them with a short series of bites with the two incisors in the lower jaw were snipped off in a few seconds. The head was held sideways so that the edges of the incisors formed an angle of about 135 degrees with the length of the tree, cutting across the fibers. The cutting surface was therefore even and oblique, but with such clear toothmarks that one could count how many bites the beaver had had to make. As soon as the tree had fallen it was dragged down to a feeding place, where it was cut up into smaller lengths by the same method used for felling (Fig. 50).

When the trees were so thick that the beavers could not bite across the trunk, they applied another technique. First they made a series of bites with the lower incisors, with the head held so that the edges formed the same angle with the stem as when they felled the smaller trees. Sometimes they tore a strip out from the wood from below upward, but usually they first turned their heads almost upside down so that the edges of the working incisors formed an angle of about 45 degrees with the trunk; they made a similar series of bites a little farther up (Fig. 54). Then the piece of wood between the two cut surfaces was torn out with a violent jerk (Fig. 57). The chips might be about four inches long and more than an inch thick if the tree was a birch, and considerably larger if it was an aspen or some other type of tree with softer wood.

A tree that was about three inches in diameter was felled in less than ten minutes, and an aspen nearly five inches through in less than half an hour, even if the animals were not working at full pressure all the time. But beyond that size the felling took longer, according to the thickness of the trunk. When working on thicker trees they often began by tearing away thin strips from below upward, and than changed over to hacking out chips downward and upward. If they could get at the trunk from all sides, they would move around it while they were working so that the gnawed surface took on an hourglass shape.

Before starting the actual work of felling they would bark the trunk around the place where they were going to gnaw. Dry, hard bark was bitten away in small pieces and flung aside, but if the bark was soft and juicy it was torn away in long strips from below upward and the pieces of bark were eaten one by one (Fig. 53). In these cases it might be a long time before they got to work with the actual felling.

As always when they were working, the two males liked to be alone when they were felling trees. The one who was first on the spot went on for as long as he could or wanted to. If his companion came anywhere near he would depart as soon as the working one started snarling. The smaller trees were felled at one go, and the one that had felled them regarded them as his property. His companion was not even allowed a taste. With larger trees they often worked in short shifts, and then they might take turns with one another. A tree that was so big that the one that had felled it could not drag it to the water provided food and work for both animals; they would cut off branches and drag them away alternately or from different parts of the crown.

It was always exciting when a rather large tree was ready to fall at any moment. It seemed as though the beaver knew he was just about to reap the reward of his labors, and he put extra force into the last few bites. Just before my own eyes had been able to register that the tree was starting to topple he would jump to one side and rush away toward the water. Presumably he was reacting to a mechanism that warned him the tree was about to fall, and as he always glanced upward while jumping it is possible that he registered the direction of fall. At any rate he always jumped in the right direction. His companion, too, rushed away to the water as the tree crashed over, and afterward they both hastened back to their own part of the crown to help themselves to the feast.

Finally we cut one of the biggest aspen trees in the neighborhood. The trunk was ten inches in diameter, and although we did not take the whole of the crown along, it was quite a bit of work driving the tree over and setting it up in the enclosure. I meant to film the felling in color, but the summer was already so far advanced that there was only sufficient light for filming a few hours in the evening and a few in the morning.

They started barking the stem immediately, but I thought that this time they really had enough to keep them occupied for a little while, and that it would be time enough to go up early next morning to film the actual felling. But later that evening I began to think about it and went back to the enclosure to see how far they had got. To my great astonishment I found the aspen already felled and the beavers munching contentedly on the lush foliage. There was nothing to be done but bring another one along the next day. We had just finished erecting it when the beavers came out and flung themselves on it with such a ravishing appetite that I dared not leave the enclosure until I had got the film I wanted.

When the beavers had barked the trunk as far as they could reach and eaten up the thick, juicy bark, one of them started the felling. He worked intensively, panting with the effort, with brief pauses to get his breath and grind the edges of his incisors (Fig. 55).

The two beavers had had only one tree that evening, and one of the beavers was now unemployed. It went impatiently up to the tree now and then to see how long it was going to be before supper would be served, but scratched himself "thoughtfully" on the nose and returned to the water when the feller started complaining. The unemployed one was hungry and he poked about in the stream after the remains of previous nights' suppers, then all of a sudden he ran resolutely up to the tree, stretched himself up on his toes so that he could reach the unbarked part of the trunk, tore off a piece of bark and scuttled back to the water (Fig. 58). It all happened so quickly that the working beaver hardly had time to react. He was beginning to get tired and after about an hour's work he went over to the stream to relax and bathe, when the other one slipped in to take his place at the tree.

After that the animals worked in shifts, changing over about every hour. They had gnawed through about two-thirds of the trunk when it grew too dark to film any more. They were growing more and more eager, and if I had not chased them away they would have felled the tree before midnight. By then they were so persistent that neither shouting nor waving one's arms had any effect on them, and in the end I had to take up my stand beside the tree with a stick in my hand.

It was a long night. As I had expected to go home first before I started watching in the enclosure, I had neither warm clothes nor coffee with me.

When at last it began to grow light I stood with the exposure meter in my hand, waiting impatiently for the indicator to give the go-ahead. At about three I was able to go and sit beside the cameras, and I had barely got them focused before one of the beavers started to work again. After four hours combined working time the tree fell into the stream with a violent crash, and all was peace and joy again after a night that had been troublesome both for the beavers and me. They were dragging and tugging at branches to their hearts' delight, and I had the pictures I wanted.

The next few nights the animals felled more big aspens. I also tried to get them to fell a trunk without any crown to it, but that trick did not come off at all. Only when I hung a branch

51 The bark is investigated before the work of felling begins.

52-53 The lowest leaves are unfortunately too high up, but the bark is juicy and good right down to the roots.

54 When the chips are bitten off from below, the head is held almost upside down.

55 The feller has to pause now and then to get his breath and sharpen the edge of his incisors.

56 It is not often possible to cut off a whole large branch while the tree is still standing.

57 When the chips have been loosened above and below they are torn out with a violent jerk.

58 The male beaver on the left quickly tearing away a piece of bark. Then he rushed away to the water before the feller had time to object.

with fresh leaves on it at the top of the trunk did the beavers decide it was worth felling.

After that the weather turned bad for as long as the nights were more or less light, and I could be thankful that I had managed to record the tree-felling at the very last moment.

It was rainy the latter part of that summer, and in August the water started rising rapidly in the beaver dam. It soon overflowed on every side, and the water rose continually. The beavers were happy as could be, while we anxiously measured the distance between the water and the top of the netting. To be on the safe side we added an extra length of netting at the lowest point, and after that we felt sure we need have no fears. But suddenly, without previous warning, the regulating authorities opened an overfull storage reservoir higher up the river. In a short time the water rose about ten feet above its normal level and the beavers swam over the net and out into the river. We never saw them again.

When the water fell a few days later, Elias and I rowed a good many miles downstream. We saw traces of the beavers in a few places, but they had gone on again and we had to give up the search. It was perhaps forgivable if we did not say anything very flattering about the river regulation as we rowed homeward against the current late that evening.

The two beavers were, of course, perfectly capable of looking after themselves when they took this opportunity of satisfying their wanderlust, but there were no great prospects of their finding what they were basically looking for. The danger was that they would wander around like hermits in ground empty of beavers, and never find the companionship that is such an important part of the beaver's life. We thought that one of them would get together with the wild female beaver whose tracks we had seen during the summer, but she was later found dead in a fish trap. If we had been able to keep them we would have seen to it that they each met a female before the autumn, and then they would certainly have regarded their area as their home.

Then we would little by little have let them colonize fresh grounds out in the open.

A few days later I happened to hear that someone had shot a seal in the river where I knew that our two males must have passed, and it struck me at once that it was one of them that had been killed. But investigation showed that it had really been a seal. The man had opened fire from a distance with a shotgun and when the lightly wounded animal had come quite close in to land he had killed it with a number of shots. But it had sunk to the bottom so that he never got his booty.

The seals that occasionally find their way up the river have to run the gauntlet of shotguns all the way. At the coast they do a certain amount of damage to fishing equipment, but in the river they do no more harm than that they catch salmon.

Most people enjoy such a rare sight as it is nowadays to see seals from the mainland, but the old ideas live on in many. Seals eat salmon and therefore one has to shoot, even if the distance is so great that there is no chance of killing them. The whole thing undoubtedly has its roots in ignorance. No one who knows anything about seals and has studied them at close quarters would be likely to shoot at them for pure devilment. And we might well grant the seals a few of the salmon that face a meaningless death as soon as they come to the first power station in the river. There the way is barred to them, and there many of them remain until the wanderlust and the urge to play have ebbed away. Some are transported on, but the playing grounds have been largely destroyed as a result of the regulation, and when the salmon make their way back toward the sea they have small hope of getting past the power station. That is the way life is on regulated salmon rivers. The salmon die without having been able to fulfill their task of providing for the survival of the species. That task is taken over more and more by man.

In the Artificial Beaver Stream

The stream aquarium at Hölle was really built for fish, and even though we had done everything in our power to ensure that the beavers would be happy there it was a very strange beaver stream on the banks of which Greta and the male Stina had to grow up.

The Salmon Breeding Institute lies in a sort of pan deep underneath the Hölleforsen power station dam in the narrow valley of Indalsälven. One goes down a steep and narrow path, and then is shut in between high sandy banks, concrete, wire fencing and roaring masses of water. The artificial setting for the violent forces of nature leaves one feeling small and oppressed.

Inside the institute are long rows of trays full of salmon spawn among which rearing batteries portion out their evil-smelling contents. Water gushes everywhere, and the roar of water grows louder still when one enters the room that contains the costly stream aquarium, which lies half a flight down at the very end of the building.

It was in such a setting that we decided that the beavers would thrive and work in the same way as beside a beaver stream in the woods. Elias shook his head as he fitted out the aquarium for the beaver experiments, and I had no very great hopes either. But I wanted to make the most of the opportunity, seeing that the fisheries men were obliging enough to let me use their facilities. And if the beavers, contrary to all expectations, were to build a dam in the aquarium I would have unique opportunities of studying their behavior while they were at work, both above and under water.

The aquarium is thirty-three feet long by six and a half feet wide, and the wall onto the observation room is glass.

Before putting in the one-year-olds, Greta and Stina, we had built a platform at the upstream end with a living box, thirty-two inches above the bottom of the aquarium. The water level was so low that the mouth of a cylinder that led up from the water to the box lay a little above the water. In addition there was a ramp direct from the channel up to the platform.

The first autumn the beavers were given a roofless living box, but they soon filled it up with building material and after that were able to hollow out a hole to live in. They transported the material up to the platform via the ramp, but they started the hollowing out from inside the cylinder. During the course of the autumn Greta and Stina built a large lodge, which took up the whole of the platform, and in this they lived for the whole of the winter and the following summer.

Before the second autumn we pulled down the lodge and replaced it with a spacious and well-insulated living box connected to the cylinder on the platform. We took away the ramp so the beavers would not be able to reach the box from the outside, for I thought it was unnecessary for them to build over their fine new box and wanted them to concentrate their energies instead on building a dam.

The bottom of the channel was covered with a thick layer of gravel and stones, and a few yards from the outlet I piled them into a ridge right across the channel so that the water was rougher there than at the outlet; the height of the water was not raised but remained at about eight inches. As beavers had been living in the aquarium a whole year the bottom was littered with barked pieces of stick and branches. The beavers had also dropped the refuse from their meals into the water.

At the beginning of September the animals built a really fine dam, just where the water eddied around the stones I had gathered together. The building material consisted of stones of various sizes and branches and pieces of wood from the bottom of the stream. They pushed pebbles, gravel and small sticks against the upstream side of the dam so that it was absolutely smooth and sloped steeply down toward a deep hollow in the layer of gravel and stones on the bottom of the channel. The downstream side came to consist of a fencework of longer sticks and branches, most of them lying in line with the direction of the current, so that they formed a support for the dam. It was absolutely watertight and the surface of the water stood so close to the crown that it needed to rise only a fraction of an inch and it would start running over. The overflow occurred only at the farther end of the dam.

The beavers, by building the dam, had provided themselves with deep water in the entire aquarium; in the channel itself the depth was twenty-seven inches, and the mouth of the cylinder leading up to the box was completely under water. The beavers, scarcely one year old, who had never seen a real beaver stream and had lived under such peculiar conditions all of their lives, had built a perfect beaver dam. They started building as soon as they were presented at the right time of year with the stimuli that are always needed for beavers to start dam-building, and the work had proceeded quickly and led without any fumblings or mistakes to a perfect result.

But the best of it all was that the beavers themselves seemed so pleased with their work. Their whole bearing was noticeably changed. Up until then they had simply eaten and slept and one seldom saw them together out in the aquarium. They grunted and snarled if they came too close to each other, and it was only inside the box that they appreciated direct contact. But now they were restlessly active in the aquarium all night. They seemed to enjoy swimming and diving in the deep water. When they ate they had nothing against sitting side by side, and at these times they "talked" to one another in the same way as Tuff and Tuss had. They sometimes sought each other out simply to "talk" or to help one another to groom its coat.

Their behavior toward each other had changed in the same

way as that of Tuff and Tuss after they isolated themselves in the second summer from the "superfluous" males. So it seemed as though Greta and the male Stina had "got engaged" now that they had built themselves a dam and thereby transformed the aquarium so that it satisfied their requirements for a home. Now they seemed completely adjusted and I was convinced that they would mate during the winter.

But for the autumn the thing was to learn as much as possible about how beavers build dams.

When the two animals had finished their dam-building they started, to my great astonishment, building over the box. I had not thought there was any possibility of their getting up to the outside of the box, but there was a heap of stones at the upstream end of the aquarium behind the actual channel, and it turned out that, if the beavers climbed up on them, they could clamber over the partition and up onto the platform above the channel. They collected great armfuls of building material on the bottom of the channel, swam with it under water to the downstream end, where they got out into the space behind the partition, and then climbed up with their loads by means of the stones onto the platform. It was a long and difficult route and it was quite remarkable that they were able to find the box from that direction. I had every reason to admire their energy and determination as they climbed on their hind legs up the uneven pile of stones with their arms full of branches, stones and shredded wood, and then with great difficulty worked their way up onto the platform. They had such a determined look in their eyes and it seemed as though nothing in the world could prevent them rebuilding their home according to their own ideas now that, at last, they had got sufficient deep water in the "stream."

First they covered the lid of the box with a thick layer, while at the same time carefully stopping up the cracks between the sides of the box, the partition and the glass wall respectively. Behind the box there was an empty space on the platform, which from the point of view of the beavers on the roof of the box looked like a deep hollow. In the autumn a beaver's building proclivities are always set going by the sight of a hollow anywhere near the lodge, and after Greta and Stina had filled up all the lesser hollows around the lodge they started dropping building material into the hole. In the end the whole area around the box looked like the smooth, conical roof of a stream lodge. The beavers then gnawed their way in from the inside of the box into the building material at the back and cut out a hollow there. As the new sleeping chamber extended right up to the glass wall, we were able after that to watch the beavers from the observation room as they lay asleep in the chamber. As a rule they were quite unconcerned about anything moving outside the glass wall, though sometimes they would come up and try to nuzzle at objects moving outside.

It was a joy to watch their activities in the water behind the glass wall. I had them right before my eyes as they worked and swam under the water, and thanks to twelve photographic lamps in a holder above the water I was able to record every movement in slow motion. I knew already that beavers spend a large part of their time under water, and after seeing the swimming demonstrations by the two males through the clear water of the beaver dam, I was also aware that they are very accomplished underwater swimmers. But even so, it was only in the stream aquarium that I really got any idea of how well adapted beavers are to life in the water.

An undisturbed beaver dives almost noiselessly. It does a gentle roll and disappears very quickly under the surface without causing any violent movement of the water (Fig. 61). As this is the quickest method it is also used when the beaver is suddenly surprised by danger, but generally the animal has time to pick up scent and movement for a moment, and these stimuli gradually induce diving with a tail beat. In this the tail is bent around over the back at the same time as both hind legs give a kick-off backward, so that the water is thrown up in a violent cascade. This part of the movement pattern lasts through seven frames, filming at the rate of twenty-four frames per second. In the course of the next frame the underside of the tail is struck hard against the surface of the water while at the same time the tail end of the back disappears under the surface, and it is this swift, hard blow that causes the dive to be so loud. As it does not contribute to speed but rather slows it down, the tail beat must have some other function, and it would seem as though the consensus that it serves as a signal to other beavers must be correct, although the beaver's ability to react to tail beats by other beavers is not inborn. Beavers often rise to the surface again after diving with a tail beat only to make a soundless dive immediately afterward, and occasionally they may do a long series of dives with tail beats one after the other, in play or if they are slightly irritated. They may also beat their tails against the surface of the water without diving, in the same way as they sometimes strike their tails against the ground if they are irritated while they are on land.

When a dry beaver dives, a string of air bubbles always rises from the spine to the surface of the water (Fig. 61). Some of the air that is present in the fur is obviously forced backward by the flow of water around the body and leaves the fur at a particular point where the rump is closest to the surface of the water. Then the coarse guard hairs close to form a sleek, silver-gleaming and—as far as can be judged—watertight sheath over the short, thick underfur.

As soon as the outer ears touch the water they fold together to close the ear opening, while at the same time the nostrils are covered with two folds of skin. The body suddenly becomes narrow and spool-shaped and makes an impression of swiftness and smoothness that contrasts strongly with the normal image of heavy, clumsy beavers on land.

Down in the water the hind legs make simultaneous swimming movements. The webbing of the feet folds together as the feet are drawn forward close under the body, and opens out again as the legs make a powerful swimming movement obliquely outward and backward. (The feet of a large beaver reach a width of six inches.) The feet thus follow a circle-shaped course and act more as propellers than as oars. The tail and the whole body make strong undulating movements on the vertical plane, and these play a large part in driving the animal forward on a slightly wave-shaped course. The downward beat of the tail has in particular a strong propulsive effect.

If the beaver is moving slowly along close to the bottom, the hind legs are stretched straight backward while the body and tail make vertical swimming movements. A beaver that is frightened while it is at the bottom may make a quick series of such movements in order to work up speed. It is almost like the swimming motions of a flatfish. The tail also functions as a very effective elevator and rudder, and it may be turned so that the flat side is standing vertically. The hands are kept still under the chin in swimming both under water and on the surface.

When the beaver swims on the surface, the hind legs make alternate strokes and the tail helps to a certain extent in that it is used as a rudder, but it is never placed vertically when it is used as a rudder in swimming on the surface (Fig. 60). The movements are not nearly so specialized as those employed under the surface, when the beaver is swimming much faster than up on the surface.

A six-month-old beaver that was put into the aquarium swam at a speed of six feet eleven inches per second under water when violently frightened. As mentioned above, it has been recorded that beavers have dived up to 2,500 feet in one dive, and at a speed of six feet six inches per second this would mean a diving time of 6.6 minutes. Frightened beavers may lie pressed against the bottom for up to fifteen minutes.

It is fascinating to study an animal's various movement patterns with the aid of film. Often one understands the functions of the movements only after very detailed study, and comparisons with corresponding movements in other types of animal sometimes give certain indications as to how the movements may have developed in the course of millions of years.

In order to find out what the beavers did when they built the dam I had to tear it down. I relied on their building it up again immediately.

When they came out in the evening and saw their work destroyed they looked worried at first, but soon they were hard at it putting things in order again. They worked with frenzied energy at first, but as soon as the flow of water over the dam began to lessen they took things more easily, and after that alternated between building and eating.

First they piled up building material of every description, and without any order, on top of the remains of the dam. They took it from the bottom of the aquarium, since gathering it together and transporting it went much more quickly there than on land. Sticks they picked up with one hand and pushed as many as they could hold into their mouths, then they shot out both hands, palms downward, along the bottom until they had a pile of sticks, stones, shredded wood, etc., between their arms and their chins, and then, clutching this firmly between their arms and chins and with the sticks in their mouths, they swam under water right up to the dam. There they put their feet on the bottom and walked on their hind legs, carrying their load up to the crown of the dam (Fig. 62). It was a remarkable sight to see them walking on two legs under water up the side of the dam and slowly rising out of the water like mermaids, with their great burdens in their arms (Fig. 63).

They dragged branches through the water and then hauled them over the crown, constantly taking fresh grips with their teeth. Sometimes the beaver would remain in the water on the upstream side and push the branch over the crown of the dam from there (Fig. 64). As it was dragged along at the animal's side the branch lay roughly at right angles to the length of the dam, and the running water also helped to a certain extent to keep it lying parallel with the direction of the stream when the animal fixed it against the bottom on the other side, holding it with its teeth and pushing energetically.

Once the dam had been built up above the surface of the water the beavers started pushing finer material against the upstream side, packing it fast with their noses and hands, but they still went on fixing all the branches they found against the bottom on the other side.

On one or two occasions the beavers were seen taking stones between both hands and lifting them up onto the dam. They rolled larger stones or pushed them along the bottom until they could get their hands in under them and clutch them firmly between arms and chin (Fig. 66). They sometimes collected sticks and shorter pieces of wood into large bundles in their mouths rather than hold them in their arms (Fig. 65).

Later on I broke the dam down on a number of occasions, and the beavers always managed to get it watertight again that same evening, provided they had access to earth on the bottom of the channel. They would use a really subtle method. They held their hands still at the bottom and at the same time struck out very quickly with each back leg alternately. This threw up mud into the water, while the current made by the violent kicking carried it toward the dam. In quite a short time the dam was completely watertight so that the water rose to the crown, and after that the beavers merely had to put a few finishing touches where the water ran over.

For the following month I changed conditions in the aquarium in various ways every day and then recorded the beavers' behavior between six o'clock and midnight every evening.

The beavers had already shown in the enclosures that they reacted to water gushing at the surface and that they built where the water flowed fastest, and in the aquarium I made the most of the knowledge to get the beavers to build in what was, from the point of view of my studies, the most suitable way. I now had opportunities for investigating the sort of stimuli to which the beavers reacted when they were building or repairing a dam.

The first problem with which I confronted them was having the upper part of the dam torn away. They started repairing this only after they had been close to the water that was rushing over the crown of the dam. If the flow of water was strong the male started the repair work at once, but if it was weaker the female might go on eating for a longer or shorter period after she had been in the vicinity of the leak before suddenly swimming out into the channel for material. After that the building would go on uninterrupted until the flow of water had been largely stopped. Then she might sit eating close to the dam, but if it was not absolutely watertight she would have no peace. She would sit listening to it the whole time, and now and then make as though to start building. She might sit like this on tenterhooks for a long time, until she would suddenly fling herself out into the channel for material and, with terrific energy, fill up the hole. It seemed as though hunger and the urge to build counteracted each other until enough stimuli from the trickling water had been stored up in her brain for the latter urge to take the upper hand and, as it were, explode into intensive building. And it seemed to be, in the first place, the sound of the running water that made the beavers start their building work.

When the dam was torn down completely, the male alone was responsible for the work of rebuilding, but if the damage was on a smaller scale both animals took part. The smaller adjustments were done by the female alone. This seems to have been not simply a peculiarity of this particular pair of beavers. The females seem to react more sensitively to stimuli from the dam, but if these become too strong the dam-building response is thrown out of gear for her, so the male has to do the heavy work alone.

Practically all the building work was done from the upstream side. The beavers did not seem to react if water trickled through the dam beneath the surface. If they ever happened to be on the downstream side when water was spurting out there, they tried to push material toward the spot, but this never produced any lasting effect.

Even when the dam was in perfect condition, and the water stood almost level with the crown and ran past only at the edges, the beavers remained fully active in the aquarium and inspected the dam and the lodge at regular intervals. They then examined their work carefully from all sides with their noses.

In the next series of experiments I made furrows in the crown of the dam. Although the water did not run so fast through the furrows and the water level did not sink so much as in the previous series of experiments, the female often started building as soon as she came close to the leakage, but if she was hungry she might sit eating for a while before suddenly starting repair work. Sometimes she would sit inside the lodge eating, after she had found out that the dam was leaking, but when in the end she came out again she always picked up building material on her way to the dam.

The animals generally filled in furrows in the crown of the dam first with shortish sticks, which they pushed into the furrow so that they lay parallel with one another and more or less at right angles to the length of the dam. After that they pushed in shredded wood and other finer material and packed it firmly in with their hands and noses. It seemed as though they even reacted to the sight of a dent in the crown of the dam, but before investigating this further I wanted to find out whether changes in the water level had any significance.

I therefore made another series of experiments, raising the water level by increasing the water supply in the channel so that the water overflowed the dam. This was intact and there were no dents in the crown of the dam that might have affected the animals. They were noticeably stimulated by the deeper water and dived and swam in the channel with the utmost delight, but as soon as they came anywhere near the dam they immediately started building on top of it if the water was flowing over the top with any force. They reacted to the running water in exactly the same way as when it was caused by the upper part of the dam being torn away so that the water level fell. The more the water rushed the more intensively the animals built, but if I put a layer of shredded wood on the crown of the dam so that the sound of the water was deadened, the beavers stopped building. It seemed as though they reacted only to the sound made by water running on the surface.

Changes in the water level seemed not to affect building activity at the dam. On the other hand, the beavers started building onto the lodge with tremendous intensity when the water rose. As soon as they had put a stop to the worst of the flooding over the dam they started on the lodge, and until the dam was completely watertight they alternated between the two building sites.

They often behaved far from rationally. After unloading an armful of building material onto the lodge they might carry material from the lodge to the dam, and after building it in there, transport material from the dam to the lodge. They could go on like this for a long time, carrying material backward and forward. They seemed unable to decide which piece of work was more important. There was a sort of short-circuiting of their building behavior, which otherwise functioned so perfectly and appropriately. This was presumably due to the fact that, when the amount of water flowing over the dam was reduced, the urge to build onto the dam was also reduced, until in the end it leveled out with the urge to build onto the lodge, which was activated by the rising water level. And as the distance between the dam and the lodge was so abnormally short, the beaver, like the famous ass, was caught between two bundles of hay.

In the next series of experiments I lowered the water level by decreasing the supply. When the water level fell, they stopped building onto the lodge immediately and concentrated all their interest on the upstream side of the dam, which was now lying partly above the water level. The same thing happened when I lowered the water level by means of a couple of syphons that sucked the water over the dam. The beavers pushed finer material against the upstream side of the dam until it was absolutely smooth, but they took no notice at all of the fact that the water was being sucked into the mouths of the syphons just beside the dam. When a leakage occurs under water the beavers evidently react to unevennesses in the upstream side of the dam, and under natural conditions the leak is closed when they plaster the slope of the dam with finer material. But only if the water level falls down to the hole, so that water starts gushing through it on the surface, do they turn their attention to repairing the actual hole.

One could often tell from looking at the beavers whether it was visual or auditory impressions that caused them to start building at the dam. As I was observing them all the time that they were solving the problems I had set them, I got quite a definite impression that it was, firstly, the sound of water running at the surface and, secondly, the visual impression of unevennesses in the crown and upstream side of the dam that determined their activities there. But I had as yet no real proof of my theory.

In the next series of experiments I first made a deep furrow in the crown of the dam, so that the water went pouring through it. I then placed a hood of plexiglass over the furrow and stuffed wood fiber carefully around it so one could not hear the water rushing underneath. One could, however, see it clearly through the glass. The crown of the dam was level with the edge of the hood; looking from the side there was no indentation to be seen in it. A little distance from the hood I made another furrow that was clear above the surface of the water. When the beavers came out they immediately filled up the dry furrow with building material, which did not, of course, result in raising the water level. They did not understand that water was running over the dam under the hood.

When the hood was replaced by a net so fine that the water could not be seen through it, but could distinctly be heard rushing, the beavers carefully covered the net with shredded wood, and, having cut off the sound of the running water, they proceeded to fill the dry furrow with sticks and plaster it over with finer material. After that they were perfectly satisfied with their handiwork, even though it had not given a normal result. After these experiments I was fairly sure that my theory was correct. While building their dam beavers react to the sound of water flowing at the surface and to unevennesses in the crown of the dam and its upstream side, if these are exposed above the surface of the water. But they do not react to water running through holes that are below the surface of the water. A rising water level seems to have no effect on dam-building, but it stimulates building on the lodge, so that consequently the lodge is adapted to the higher water level. When the water level falls the beavers grow uneasy, stop building onto the lodge and swim restlessly up and down, so that they soon come in contact with the disturbances at the dam that stimulate dam-building.

By that time I had spent a month at Hölle and interesting and delightful as the beavers were I thought it would be pleasant to turn to other tasks for a time. If the beavers were agreeable I thought of resuming the experiments later on in the winter when I hoped to find out how mating took place, and then, in the early summer it would be really exciting to see how Greta brought up her young in the aquarium.

But it was not long before I again received bad news from Hölle. Greta had gnawed a hole in the bottom of the box and got wedged under the platform so that she drowned. Stina completely changed his way of life and pined all through the winter.

The End of the Story and the Beginning of Another One

I have now told the story of how I was able to make acquaintance with the beaver, and of the individuals who gave me my first insight into the world of beavers. I have not as a rule had time to become so well acquainted with the beavers I have had charge of since for longer or shorter periods, with the exception of the male Findus, who in spite of the fact that he is now mature refuses to have anything to do with other beavers and insists on regarding me as his life's companion.

After the first three years of experiments with beavers I had every reason to be satisfied. The beavers had demonstrated a large number of movement patterns that are characteristic of their species, and presumably it was only the mating I had not yet seen. As I had also been studying wild beavers both out in the open and in captivity, I knew that the tame beavers had for the most part really demonstrated the normal behavior of the species. An animal's various movement patterns are just as characteristic of the species as are, for example, the shape of the various physical structures, so that an account of the behavior of a species is valuable not only in itself, but can also serve as a basis for comparison with accounts of other species. If one wants to form an idea of how any species of animal is adjusted to its environment, it is obviously necessary to know in detail the various movement patterns and instinctive actions of that species. It is often particularly instructive to study how these are developed in the young. And only after we had got to know the behavior of the beaver had we any possibility of judging the effect of the short-term regulation on the beaver colonies.

My main concern, however, was to study the relationship between what was acquired and what was inherited in the beaver's behavior, and in this respect I seemed to have found an ideal research subject.

The young beavers I had brought up on a bottle had developed normally, in spite of their unusual surroundings. Everything seemed to indicate that the beavers' movements and their co-ordination into complicated movement patterns are inborn. At all events the young ones need not learn them from their parents. But in order to analyze the problem more closely I had to bring up more young ones, and keep them isolated from those factors in their environment to which the various forms of behavior are adapted.

To try to get hold of newborn beavers out in the open is a very chancy undertaking, so I hoped that Tuff would have several young ones the next time and that I could adopt one or more of them.

It is a great moment every spring when one first sees the beavers up on land, but this spring it was particularly exciting. All the beavers had now had time to adapt themselves and were very much at home, each in its own territory and with its own family, and in May I was able to take pleasure in seeing a particularly active beaver life on the shores at night. Tuff was, quite correctly, pregnant for the second time and the X-ray showed two fetuses. The whole family—Tuff, Tuss and the one-year-old—was then moved back into the terrarium.

Three days later Tuff seemed uneasy and I started watching her carefully. She soon became more anxious for my company than ever, and she started complaining, with pitiful, whining contact sounds. It was clear that she was growing worse and worse. In the end she lay pressed close against me, looking up at me as though she thought I would be able to help her in her distress. I realized then that she was mortally ill, and the hours dragged by endlessly slowly. I was in despair.

If Tuff died it would mean a serious break in my work, but that was not the main reason why I was so deeply involved and distressed. For three years we had regarded her almost as a member of the family, even though she had lived a large part of the time as a wild beaver out in the open. It was Tuff more than any of the others that had been the connecting link between us and the world of beavers. I had had a number of dogs in my life to which I had been very attached and which I had mourned as personal friends when they passed away, but Tuff had given me infinitely more than had any of my dogs. I would never have believed that one could become so deeply emotionally attached to an animal as I now discovered I was to Tuff.

She died after a long day of severe pain. I opened her immediately, but both fetuses were already dead. One of them had been dead for several days and Tuff had obviously died of septic poisoning.

Tuss and the one-year-old were put back in the enclosure, and the relationship between father and child proceeded to develop in such a way that I began to suspect that the latter was a female. The next year the assumption proved to have been correct. One day we found her dead in the enclosure and she proved to have been pregnant with three fetuses. She had obviously died in the same mysterious way as her mother the year before. The post-mortem showed an infectious disease that presumably occurs even in beavers out in the open.

Tuss was a widower for the second time. He ate nothing for more than a week and simply roamed restlessly up and down in his enclosure, but whenever he happened to find himself outside the netting he worked intensively, as before, to get in again. Later on I tried time after time, without success, to give him another female. I was beginning to think he was determined to live as a widower in his old home for the rest of his life, but he has now at last been "taken in hand" by a large, powerful beaver matron.

Depressing as Tuff's death was, I never lost my optimism with regard to the possibility of continuing my beaver studies. I decided to devote the early summer entirely to catching beaver babies out in the beaver grounds, though it was more distasteful than ever to dig into the beaver lodges.

When at last I managed to get hold of a sufficient number of little ones, they quickly died of diarrhea. It was only later that I found out the composition of beaver milk. It is very rich in fat and protein, but extremely poor in sugar. As human milk, on the contrary, is particularly rich in sugar, it was hardly surprising that the young beavers' stomachs were upset.

The only result of my efforts to obtain newborn beavers that year was Findus. As he was only three to four days old when caught, he became very devoted and hungered constantly for company and attention. If we devoted too little time to him, he stopped eating. Then I was obliged to groom him to get him to brighten up again. I pinched his fur between my nails and tried to imitate the behavior of the female beaver as far as I could. Before very long he started grooming me on my hands, but as I had no thick underfur as protection, my hands were soon covered with blood as he worked them over with his knife-sharp teeth. It could not be helped. He enjoyed the social grooming in company with his keeper very much, and it seemed that it satisfied a need in him that was just as primary as eating. I am quite convinced that it was during these intensive grooming sessions, which seemed to be as enjoyable to the youngster as they were painful to me, that the strong and lasting bonds were forged between us.

The task of catching beavers on Faxälven had become more and more trying. The valley had lost most of its charm since river regulation had been brought in, but with the exception of the stretch of river between Räbbstuguforsen and Vagnforsen, which was completely devastated, conditions for beavers had improved since the two first desolating winters of regulation. Variations in the water level were no longer so violent that the animals' winter stores were washed away, and the dams formed by the regulation ice were, on the whole, no longer so high that the beaver lodges were flooded out. Beavers had a great capacity for coping with difficult conditions even in wintertime, and it seemed as though the small remainder of the colony that still persisted was increasing. In the autumn of 1961 there were still some forty animals along the river, even though we had caught a total of ninety-one.

We grew more and more doubtful whether we were doing the right thing in continuing to catch them, but the regulation authorities threatened every year to regulate more severely than ever the following winter, since they were anxious for us, as they expressed it, to "clean up" the beavers. The Riparian Court decided, however, in the spring of 1962 that the beavers should not be moved away from Faxälven if there was any possibility of saving them in any other way. We hope now that the river will become a beaver river again, and thereby recover something of the charm it lost when short-term regulation was introduced.

Family Life

However fantastic the first recorded tales of beavers of earlier days may be, they seem to me now pretty pale in comparison with reality.

The beaver's skill as a builder is remarkable enough, but what made the deepest impression on me is beyond question the beaver's family life.

We have a long way to go before we can give anything like a complete picture of the beaver's social organization, but due to my observations in the terrarium and enclosures, coupled with observations by myself and others in the beaver grounds, I can now begin to piece together the main outlines. In any case I have seen enough to convince myself that, as regards family life, the beaver is a very interesting animal. It is not the least of its characteristics that makes it a unique phenomenon in the animal world.

Young beavers often remain in their parents' homes until the spring in which they complete their second year, even though they are sexually mature the previous year. They are then far from fully grown (weight eighteen to thirty pounds), but they may already enter into a possibly lifelong alliance with an individual of the opposite sex. I do not know how usual it is out in the open for youngsters of one year old to form pairs, but in enclosures and terrariums all the beavers have "got engaged" during their second summer if they had the opportunity to do so, and after that they have built themselves a home of their own, mated under the ice in February and had their first litter of young in the spring in which they completed their second year.

The young of the year before may remain in their parents'

home even during the spring and summer in which the mother gives birth to more young ones, but in any case later that summer the one-year-olds are no longer so strongly attached to the parental home and they often set off on journeys of exploration that grow longer and longer during the course of the summer. It would seem quite natural if they were then seeking themselves a partner, but exactly what the young beavers do that summer is one of the most mysterious things about them.

Some of them return to their parents' homes early in the autumn, and if they do they take part in the family's autumn occupations and live with their parents through the winter. But the following spring they are seized by extreme restlessness, which drives them away from their parents' homes. They go out into life of their own free will and are not driven out by their parents, as was formerly believed—certainly not if the family has plenty of food. If they should wish to return home when their wanderlust has been satisfied by the summer's rovings, they seem to be welcome, but more usually they build themselves homes in the autumn, together with partners they may have met during the summer.

Some of the young beavers, then, stay at home until they are two, while others take themselves off as one-year-olds. If food is short in the territory the members of the family get a little irritated with one another, particularly if the family is a large one. Irritation and the competition for food certainly may play their part in loosening the bonds between the youngsters and home, but the need to find partners also probably enters in.

We know that brothers and sisters often pair off together and that young beavers that have been introduced into a group in the course of the previous summer prefer the individual of the opposite sex with which they have been together longest when they pair off a year later. Furthermore it seems that a male that has become a widower prefers to join up with a daughter rather than with a strange female beaver. It would therefore not be surprising if young beavers out in the wild pair off with brothers and sisters if they have the opportunity to do so. But even if pairing between brothers and sisters is normal, the youngsters must often have to look for a partner outside the home, since the number of babies in a European beaver litter is never more than four and in the Faxälven district seldom exceeds two.

It is possible that brothers and sisters that are able to pair off leave home to found a home of their own, while the others stay on another year before setting out on the hazardous undertaking it must be to wander around in the world without a life companion. Beavers are not fully grown until the autumn following their second spring (weight thirty-five to fifty-five pounds).

As I have already said, the Faxälven beavers, when first caught, were put together two by two in boxes until the male and female had got to know each other well enough to sleep together, groom one another and so on. Generally all went well if a male and a female were put into surroundings that were strange to them both, but best results were obtained from putting a large female together with a small male. If both animals were the same size, all usually ended happily, but if the male was very much bigger it could even happen that he would kill the female. If the male was at home in the box before the female was put in, there were generally complications, and in any case it was a long time before the animals made friends. These conclusions were possible after I was able to witness pairing procedure in terrariums and enclosures on seven different occasions.

I have also told about when Tuss came back to Tuff after a period of grass widowerhood. Tuff showed quite clearly that she recognized her mate, but she thought it best, all the same, to give him a short refresher course in pairing procedure, to be sure that he would remember his place as submissive husband. After that, the relationship between Tuff and Tuss grew warmer than ever.

I had a particularly good introduction to pairing when I

59 In the summertime the beavers lead quite a lazy life.

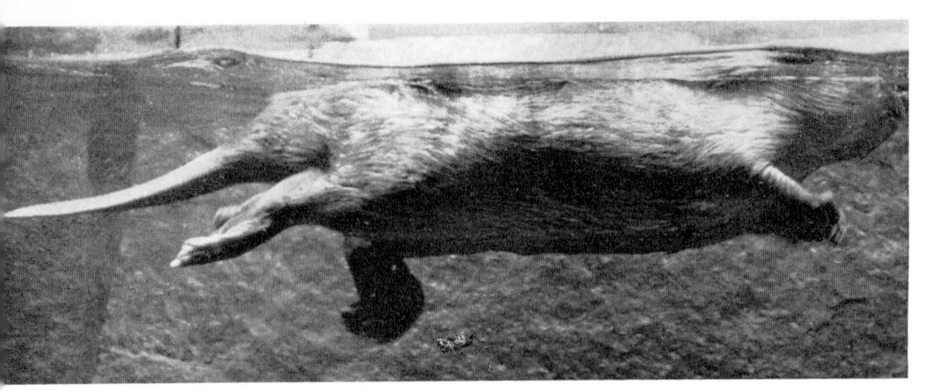

60 At the surface of the water the beaver swims with alternate movements of the hind legs. The hands are held still under the chin.

61 An undisturbed beaver dives almost soundlessly. Both hind legs are drawn forward and then make simultaneous swimming movements when the beaver swims under the surface.

62-63 Beavers like to collect their building material at the bottom of the watercourse and transport it under water all the way to the dam. There they stand up and walk on their hind legs up the upstream side of the dam to dump their load on the crown.

64 Trunks and branches are hauled over the dam to be fixed against the bottom on the other side later.

65 Beavers sometimes carry big bundles of sticks in their mouths. The fish in the background often sought the shade under the beavers' stomachs.

66 A rock is worked toward the dam with the aid of hands and chin.

67 Mother, father, one youngster from the year before and two of this year's four eating together at the feeding place outside the lodge.

68 At regular intervals the adult beavers go to the boundary of the next territory to "scold" at their neighbors for a little while.

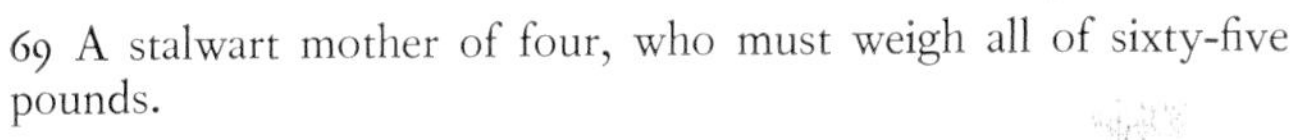

69 A stalwart mother of four, who must weigh all of sixty-five pounds.

70 A little petting for the youngest.

71 Findus building on his box at the age of forty-five days.

introduced the male Findus at the age of a year and a half to a female of the same age. Findus had been together for a time with a smaller, wild male that had difficulty in adjusting himself to captivity, partly because Findus looked to me all the time to satisfy his need for contact. He dominated his beaver companion and developed an ugly tendency to bully members of his own kind.

The two males were separated in the autumn and each was given a female that had grown up at Skansen. As the females were accustomed to captivity, they did not seem at all put out by finding themselves in a strange place. The one that was put in with the smaller male immediately took command and, because of that, the two animals were good friends after the first twenty-four hours.

Findus, however, was quite beside himself when he saw the strange female, and partly due to the fact that he was on home ground he succeeded in a very short time in creating such terror in her that she ran away as soon as he came anywhere near. After two nights there was nothing to be done but take him away. The female immediately took courage, inspected the whole area carefully and then devoted most of her time to depositing castoreum all over the place. When, two days later, I put Findus back with her, her whole attitude was completely changed. When Findus came rushing at her she calmly stood her ground and gave him the reception he deserved.

After a few violent wrestling matches Findus had to admit defeat by the considerably smaller female, but she had to fight hard for several nights before she could feel she had the fat, spoiled Findus properly in his place. She soon established special places on which she carefully deposited castoreum between battles, obviously to convince herself that it was she who was in command. When all might normally have changed into a more loving relationship, Findus unfortunately refused to accept the female's approaches and went on addressing all his affection toward me.

The depositing of castoreum obviously plays an important part in pairing, while at the same time has significance in the marking of territory. It surprised me for a long time that strange beavers are not repelled by the markings of the owners of the territory. A beaver that has penetrated into a strange territory is, on the contrary, eager to seek out the castoreum places in order to deposit his own castoreum there.

I am beginning now to be convinced that castoreum has an indirect significance. When the members of a family know their own and each other's markings they feel quite simply at home, and are stimulated to defend their territory against strange beavers. When a female wants to attract a male, she first deposits castoreum inside her future territory, which gives her the stimulus she needs to get quickly the necessary domination over her intended partner.

After pairing, the male, too, starts marking the castoreum places, but only after the female has just covered them with fresh castoreum.

It is presumably usual for the females to set up their territory first, to which intending suitors are attracted, but certain observations indicate that the female may also pay court to the male.

When the young females set off on a courting journey, they obviously have a better chance of acquiring a mate if they are big and strong. If beavers have no partner when they leave the home, they have to accomplish both pairing and home-building before the winter, and to achieve that they most certainly need to be fully grown. The young ones that pair off at home in the spring and early summer have, on the other hand, the whole summer and autumn in which to make their own homes. It would seem to be important for the species that the younger ones spread out and colonize fresh areas as early as possible, but at the same time it is important that the animals that have to go out on long journeys to find themselves a partner should be fully grown.

Family groups are most successful in adjusting themselves to their environment if they are reasonably large. A good-sized labor force is needed for the comprehensive work on dams, lodges and felling sites, and it is warmer in the lodges in wintertime if a number of beavers live together. Beavers are so sociable by nature that they quite simply are happier in groups, and observations in Russian beaver farms actually indicate that the female beaver has more young if the family group is fairly large. We know that the size of the groups and the number of young is dependent largely on the supply of suitable food. It is conceivable that this is due partly to the fact that the groups are able to keep together all the year around, if there is plenty of food, and that the females then have more young because they are stimulated by the social life with the other members of the group. If this is so, it would explain why indiscriminate hunting has such a fatal effect on beaver populations. The extermination of the beaver from large areas of the world came about astonishingly quickly. On a journey through the American beaver grounds I got a strong impression that it is only in those areas where there is no hunting that the beaver groups are really vigorous. Yet hunting is not done nowadays on any large scale. The only areas in which this occurred have been fully protected now for a number of years. If my theory is correct, hunting should be reorganized in such a way that certain beaver groups would be left completely undisturbed.

Groups must not, of course, become so large that they use up more food than can be replaced by annual growth, but there is no need to worry about controlling the number of individuals within the various groups, as this seems to be regulated automatically in relation to the food supply in the group's territory.

It is not only the risk of a food shortage that makes it desirable for the group not to grow beyond a certain limit. Beavers never collaborate directly in the search for food or the work on lodges, dams and winter stores, and if there are too many animals in one group they get in each other's way and have to compete for food and working places. The members of the group then get increasingly aggressive toward one another, and this leads to the young ones leaving home. Beaver groups therefore never become larger than ten to fourteen animals, no matter how much food there may be in the territory.

Presumably, if the family is small, the young ones remain at home, so that the group quickly reaches the optimal size, but if food supplies are short, they may leave the family before it has reached maximal size. The male may even leave the territory during the summer so that the female and the youngsters that remain in the lodge can get sufficient food. The birth rate then drops—possibly simply because the female is not happy left alone through the summer—the vegetation has time to recover, the male gradually ceases to have to live as a grass widower through the summer, and the young ones are able to stay at home again.

However beaver society may be organized in detail, it is clear that it is based on stable family groups, the size of which is sensitively adjusted to food supplies and the demands made by the beaver's highly specialized way of life.

Beavers can grow to be at least twenty years old and beaver homes are generally quite stable structures. In the poorer beaver grounds, of course, it happens not infrequently that a pair of beavers has to move on account of lack of food, and this is one of the great tragedies out in the beaver grounds. A pair of beavers become so strongly attached to their home territory that only the utmost necessity can force it to leave. But the tragedy is even greater if the male and female for some reason or another have to part.

The relationship between the two animals in a pair is without doubt the most striking thing about the beaver's behavior. Once the often somewhat violent pairing ceremonies are over, there is never any trace of discord to be seen between the partners. They sleep curled up close together during the daytime, and at night they seek each other out at regular intervals to groom one another or just simply to sit close side by side and "talk" for a little while in special contact sounds, the tones and nuances of which seem to a human expressive of nothing but intimacy and affection. And this relationship may be entered into at the age of one, more than six months before the animals come into heat for the first time. Since they normally live in close contact with one another all the year around, no complicated preliminaries are necessary for mating to come about. The scent of the female in heat is presumably enough to make the male sufficiently sexually stimulated, and when she goes out into the water in a particular way he follows and mating takes place stomach to stomach, the animals swimming slowly forward.

On Russian beaver farms the actual mating has been observed, and it has also been possible to determine the length of pregnancy.

The young are born after a gestation period of 105 days, and the male remains with the female in the living chamber during the birth. I was therefore taking a quite unnecessary precaution when I separated Tuss from Tuff when she was going to have her young. Professor Hediger of Zürich Zoo once observed an actual birth, and the most remarkable observation he made was that the male and female ate up the placenta together.

The male then shares in the care of the young. The youngsters from the year before like to be present in the nursery, and are often very interested in their small brothers and sisters. Family life in a beaver lodge in the early summer is a real idyll, and I never tired of watching a family in one of the enclosures, where the female had four young ones and where all seven members of the family lived in the greatest harmony (Figs. 67, 68, 69, and 70).

For the first twenty-four hours after delivery the female remains with the young ones in the sleeping chamber, but after that she goes out to get food and returns to eat it beside the pool in the feeding chamber for a time every night. After about four days the young ones go to the lodge's pool for the first time to bathe, but after a while the female carries them back to the sleeping chamber. After that the young ones bathe for a time every night, and the parents have no difficulty in keeping watch over them, as they are still shut up in the lodge. But when, at about ten days old, they start diving it is more difficult to keep them indoors. In their swimming expeditions under the surface they are more and more likely to lose their way in the water-filled passages out into the open, but one or other of the parents always brings them back. The excursions grow longer and longer, and when at the end of about a month the young ones are sufficiently developed to be able to accompany their parents on short expeditions out into the open, they are quite familiar with the way out. Normally, however, they never leave the lodge until they are two months old. It is only then that they are capable of looking after themselves and collecting their own food outside the lodge. By the autumn they are ready to take part in the work on the family lodge, the dams and the winter store.

That young beavers are so extremely well developed when they are born and yet are looked after so carefully by their parents is certainly related to their life by the water. If they were not so hardy and did not have water-repellent fur they could easily be numbed with cold when they fell into the water, and it is important for them to start swimming as soon as possible so that they have time to learn the passages to the lodge and the territory outside the lodge before autumn, when all members of the family have to devote all their strength to preparations for winter. Due to the parents' intensive supervision, they can start swimming at four to five days old.

In the spring it sometimes happens that beaver lodges are flooded out by the spring floods, while the babies are small. Tuff demonstrated in the terrarium that on such occasions the female can dive out of the lodge with one baby at a time and gather them all together in some suitable temporary hollow, where the parents' remarkable ability to keep their young together is a decided asset.

It is in this intimate contact between the young ones and the other members of the family that strong and lasting bonds are forged.

The babies are very lively and get up to all sorts of mischief, such as normally would be classified under the heading of play.

We do not know what the function of play is. Probably it has different uses for different species. Among rodents "play" is a comparatively rare phenomenon. It seems to be most highly developed in the beaver, which also occupies a special position in the matter of social behavior. The play of young beavers seems to consist of elements important to their social life, and it is therefore natural to assume that it fulfills some sort of function connected with their social "education."

Findus has shown that beavers normally learn the correct way of reacting to other beavers in their daily life with the other members of the family. As he had no opportunity of learning that he was of the beaver species while he was still receptive to such learning, he continues to regard me as a beaver and his proper beaver companion as a stranger. He is, as they say, imprinted by his keeper. That is, he has come to regard himself as being of the same species as his keeper. The imprinting seems to take place during a relatively short, critical period with most animals, including man, but it is of course most pronounced in social species.

My experiences with the fourteen young beavers of which I have had charge for longer or shorter periods indicates that young beavers are imprinted during the first months of their lives, and there seems to be a certain connection between the ability to be imprinted and the development of flight behavior.

For the first week the young do not appear at all frightened and have only a few simple protective reactions. If they are caught during this period they can become completely imprinted by their keeper. After that, the protective reactions that beavers show when they are in or near water begin to make themselves apparent, and these are fully developed when the young are about a month old. At that age they still appear quite unafraid if surprised far from the water or inside the lodge, and young ones caught not later than that age learn, like Tuff and Tuss, to recognize their keeper at a short distance and in certain situations. Indoors they react to their keepers as they do to members of their own species inside the group, but out in the open they behave as completely wild beavers. When the young are about two months old all the flight reactions are fully developed, so that they appear frightened even when they are on land, and after that they can never be imprinted by human beings.

The females seem to have greater potentialities than the male for developing lasting contacts with other creatures, and the period during which they are receptive to imprinting seems to last longer with them than with the males, even though the males can obviously become greatly attached to human beings if caught sufficiently early.

The mutual grooming, contact sounds and contact body-to-body and nose-to-nose are the most important means of social contact among beavers, except for the movement patterns, which also occur in the play behavior of the young. Apart from that, beavers have very expressive eyes. It is very easy to see what mood a beaver is in at a given moment, and I presume that other members of the group can also see this.

The most intimate family life takes place in general inside the lodge, and anyone outside has no chance of getting an insight into this under normal conditions. Outside the lodge beavers never even "talk" to one another. They use instead the special contact sounds, which have such a high frequency that not everyone can hear them.

The highly developed social behavior is interesting, and doubly so because the beaver's behavior otherwise shows many primitive features that are perhaps not common among the higher mammals.

"Beaver Pedagogy"

When young beavers start disporting themselves outside the lodge they weigh only about seven pounds. During the "nursery period" of about two months they have had no opportunity of making acquaintance with most of the phenomena which, from their first autumn onward, will play such a large part in their existence, and they receive no instruction from their parents, not even when they suddenly emerge into the new and dangerous world outside the lodge.

Our beavers have shown that it is principally only the social behavior that is dependent for its normal development on contact with other members of the group.

That other behavior patterns are developed independently of contact with other beavers is not surprising in view of what we know of the behavior of other rodents.

Young beavers can walk as soon as they are born, and after a few hours they can hiss and fling themselves to the side if some enemy tries to catch them inside the lodge. But the flight reactions that come into use outside the home are not developed until toward the end of the nursery period. They are generally fully developed when the young leave the lodge for the first time at the age of about two months. Otherwise the first encounter with the dangers out in the open might easily have disastrous results.

Tuff's daughter grew more and more timid as the flight behavior matured, in spite of the fact that I handled her every day and her mother was on very intimate terms with me. At the age of two months the youngster was more timid then wild beavers in captivity, but after that she grew by degrees as half-tame as are beavers that have been caught as adults, and about as quickly. Even out in the open the one-year-olds are very timid when they are outside the lodge, and that is why one so seldom manages to see them during their first autumn. Little by little they begin to learn not to react to movements, scents and sounds not of a dangerous nature so that the flight reaction is adapted to conditions prevailing in the home territory. The development of the flight reaction to a form that is of practical use to the small beavers is evidently controlled chiefly by inherited tendencies, and only after the behavior is fully developed is it modified in terms of different surroundings. But it is only the reaction to the various stimuli that is modified, not the movement patterns and their co-ordination.

It is easy to understand the advantages of young beavers not having to learn by experience to react in the right way to dangers, but what about those forms of behavior that they have the opportunity to practice in peace and quiet in the security of the lodge? Do they depend on practice and experience to reach their perfect form, or is their development determined to an equally high degree by inherited tendencies?

The young have a great deal of trouble with their toilet in the lodge. At first they seem very clumsy when they wash themselves, but they seem to make progress every day, and by the time the nursery stage is over the grooming behavior is fully developed. It seems as though constant practice plays a large part; but the first time these movements are called into play, after the first contact with water, they are fully formed and co-ordinated. The reason the young ones appear clumsy at first is because the ability to adopt the correct physical positions develops more slowly than the grooming movements. And the other movement patterns, too, which come into play in the nursery period, seem also to be inborn.

Findus, for example, had no access to solid objects for the first two months, but then, when he was allowed to bark pieces of aspen wood, he handled them just as skillfully from the start as any other youngster that had fumbled with pieces of wood for a long time before the behavior pattern was fully developed. And Findus swam like an adult beaver the first time he was put into the water, at a stage when the swimming movements had only just matured in other youngsters that had been splashing about in the water since they were four days old.

But the most interesting sides of the beaver's behavior—the building activity, the territorial behavior, the collecting of stores, etc.—do not come into use until the youngsters' first autumn, after they have had a whole month in which to familiarize themselves with the various peculiarities of the territory. Surely one may assume that the real learning period begins when they leave the lodge for the first time?

The digging movements are perfect even when quite small beavers that have never seen earth try them for the first time, and when Tuff dug on the floor in the terrarium she demonstrated that even the sequence of the movements is inborn. That this is also the case as regards building behavior was demonstrated by Findus at the age of barely a month.

As soon as all the youngsters we had kept isolated indoors had the opportunity to build in their first autumn they built like normal adult beavers, apart from the fact that they took longer to get their lodge finished. As the body weight of one-year-olds in the autumn is only about a quarter of that of their parents, they have not, of course, the same physical strength; and young ones that for some reason are left to themselves out in the open have little hope of getting their lodge finished and collecting a store before winter.

When Findus was forty-five days old he had never seen a stick or any other kind of building material. The only addition he had had to his milk diet had been a few bits of green leaves. He had a little box on the floor in our home and a little bowl to bathe in. One day I put a handful of leaves into the water, and to my astonishment Findus immediately started lifting them up onto his box. He used exactly the same movements as adult beavers when they build: pushing the leaves together into an armful, lifting it up and pressing it down on top of the box. When he had used up all the abnormal building material in the bowl he went on building in the empty air, still performing the movements in the right sequence (Fig. 71).

In the autumn, when it was time to start building, we moved Findus down into the cellar, where we gave him proper building material and a box like a block of salt, which was so high that he could not reach to put material on the top. He had already shown that the urge to lift material onto the top of the dwelling house is inborn. Now I wanted to see whether he would build against the smooth, regular sides of the box, and if so, how he would set about it.

As soon as he had settled down in the box he started filling up the entrance hole with building material, as all the other beavers have done that have lived in our terrariums, and after that he had a frightful job pulling the material out every time he wanted to get inside. But he built just as often against pieces of black paper that I stuck here and there on the sides of the box. When I fitted the entrance with a long entrance cylinder he stopped "closing the door after himself," and when a little later I removed the cylinder he no longer built up the original entrance to his home. He had obviously accustomed himself to react to it as to a dark hole in the wall of the house, once he had passed through for a time via the cylinder. After he had been building for a time he also ceased to react to my pieces of black paper, but he was particularly sensitive to any unevennesses on the route he used over the lodge in carrying up material. So he had learned not to react to a number of those stimuli to which he had an inborn capacity to react; possibly he had learned to react to certain new stimuli. As I said before, the young learn to concentrate their building activities on the home as soon as they begin to "feel at home" there. But the actual building patterns and their sequence are not changed during the course of the building.

The behavior pattern is relatively simple and is deeply embedded in the mass of inherited urges. Yet it can be adapted to different situations, so the result is always a marvel of suitability. In the course of millions of years nature has envolved a building method that is really first-class.

Many of the beaver's behavior patterns are common to all or most of the various groups of rodents and presumably also occurred in the unknown ancestors of the beaver. That these patterns gradually were specialized in life in and beside the water is presumably because this provided effective protection against enemies, but possibly also because it was easier to transport branches there. We may doubtless assume that bark was a favorite food even among these hypothetical ancestors.

Most rodents are skillful diggers, and a hole in the ground is undoubtedly the original rodent home. To the animals that were gradually to develop into beavers it must have been natural to dig themselves into riverbanks, and the need to protect themselves against enemies may have led quickly to their placing the opening to their dwellings under the surface of the water. In modern young beavers the digging behavior is fully developed even in the early nursery period, and this may perhaps be interpreted as an indication that this behavior is very old.

The building movements, too, are fully developed quite early in young beavers, and the ability to isolate themselves from enemies and cold by hollowing out a dwelling place in the heaped-up material is also common behavior among rodents. It is probable that the beaver's ancestors began at quite an early stage in the development of the species to strengthen the roof over their dwelling holes with the same sort of material that modern beavers use.

After the openings to the system of passages had been placed under water, the passages above water level reached a point in the shore where they were often eroded by running water, so that holes often occurred in the roof of the passages. If the shore was low the animals were forced at high water to dig passages so close to the surface of the ground that they often fell in.

Even rodents that do not collect stores often take food to their homes or close to them, where they usually go to eat.

Young beavers do this too, so it is natural to suppose that this behavior also occurred among the beaver's ancestors. Barked pieces of wood and other refuse would then collect on the shore above the passages, and the animals possibly pushed them into heaps in the same way as small beavers still do. The ability to react to the stimuli from holes in the roofs of passages by collecting refuse over them was favored by natural selection, and after that the capacity was no doubt also developed to react to moisture inside the passages and to rising water levels. On steep shores the result was an effective protection against erosion, and on low shores it enabled the animal to hollow out a dry dwelling place in the heap of material. Natural selection may also have contributed to the development of the behavior used in building stream lodges and river lodges.

Beavers put a great deal of energy into making all their passages, tunnels and other paths smooth and even, and similar behavior occurs among many other rodents. When beavers plaster their lodges with mud from the bottom of the watercourse they always build a smooth and even path with the plastering material from the water up to the lodge, and when they eventually reach the top of it they start all over again from another point on the shore. They go on like this until the whole lodge has been covered with an even layer. As soon as the pile of branches and pieces of wood is so high that they have difficulty in carrying any more material up, they simply build transport roads of finer material. This behavior has been favored by natural selection over millions of years by reason of the fact that the lodge became more resistant to erosion and against enemies. A lodge that has been built by the beavers of today is so strong that no enemies except man can dig their way through it.

But when a beaver climbs clumsily up a lodge on two legs with a great armful of clay clutched against him, this is the great opportunity for any enemy that may be lying in wait, and the opportunity is greater the farther the beaver is away from the water. Carrying material around to the back of the lodge is so dangerous that the risks outweigh the advantages of plastering, more particularly as some part of the lodge wall has to be permeable to air. The practice of leaving unplastered that part of the lodge that is hardest to reach from the water has therefore been favored by natural selection.

But it is more than anything his ability to build dams that has distinguished the beaver. Dams are the most impressive results of his activity and they are perfectly adjusted to all sorts of different environments. If young beavers need to learn anything at all, it should be in connection with dam-building.

In their first autumn young beavers join in with the rest of the family in the work of building the dam. But they seem not to be able to build a dam by themselves until their second autumn. Then they build perfect dams as soon as they come in contact with the necessary stimuli, even if they have not had an opportunity to practice the autumn before.

On the whole beavers apply the same movement patterns when they build dams as when they build lodges, and they react to the same sort of stimuli in both cases. The force of the running water contributes to a certain extent to the fact that the movements achieve a different result in dam-building, and the sound of the running water is the most important stimulating and guiding urge. It is easy to suppose that the dam-building behavior has developed out of the lodge-building behavior. The beaver's ability to react to the sound of running water by building was favored by natural selection in that the behavior led to a higher water level, which gave better protection against enemies and better waterways.

Beavers can sometimes live for a long time in a place before they start building dams, but if the water ripples between, for example, a few rocks, the dam-building behavior will normally be released even in the first autumn. If the bed of the stream consists of earth, the beavers will push it up in the direction of the releasing stimulus, and go on building where the water pours over the barrier they have pushed up. They use whatever material they happen to find when the urge to build comes upon them. Because all the material is floated up, it lands in different parts of the dam: long branches at the back as support, finer material as plaster on the upstream side, and so on. On stony beds the beavers push smaller stones together to make a barrier and then carry larger stones up onto the crown so that they roll down and form a support on the downstream side. And if the stones are all so large that the beavers cannot lift them they fill up the channel of the stream with whole trees. Since they get hold of these by the root end and float them with the current to the building position, they end up parallel with the direction of the current and with the crowns pointing upstream.

That it is primarily the sound of the water pouring through the dam that directs the building was demonstrated particularly clearly by the experienced wild beavers in one of the enclosures, where we led the water through a pipe that lay at an angle underneath the center of the artificial dam. They heard the water rushing deep underneath the crown of the dam and so they built there, high above the surface of the water. The result was a pile of building material that looked more like a lodge than a dam, and the top of which was exactly above the source of the sound. The reason the crown of the dam is usually flat and lies just above the surface of the water is that the beavers build until they can no longer hear the sound of water pouring over.

And it is only certain characteristics of the sound to which beavers react. I am at present carrying out experiments to try to find out what these characteristics are. Four different groups of beavers build industriously to a loudspeaker if it plays the right sounds. The recorded sound of running water, of course, always starts them building, but so does the sound of an electric razor. But they take no notice at all of a constant tone no matter what frequency. By filtering out various frequencies from the sound of running water we have found that the beavers react to oscillating sounds of certain frequencies, but I hope to be able to isolate the activating stimulus more precisely. It is possible that the choice of material is determined to a certain extent by the frequency of the sound.

If there is no place in a beaver territory where the water rushes in such a way as to provide the activating stimulus it may be a long time before the beavers start dam-building, but sooner or later the correct stimulus will be provided as a by-product of their other activities. They carry stripped sticks and other refuse into the water outside the lodge, and they dig away material from the bed of the stream when they deepen the passages they use when they are swimming under water. The material is carried with the current a little way before it sinks to the bottom, and as soon as enough has collected at any point to make the water begin to ripple the beavers start carrying all their refuse there. When the dam-building behavior is activated in the autumn the refuse heap is soon converted into a dam.

Generally, but not always, the building leads to an appropriate result. In the enclosures even the experienced animals always started building on the inlet side of the stream, but as the building gave no result the activating stimuli gradually lost their effect, and after that the water could make as much noise as possible at the inlet without the animals taking the slightest notice. Beavers can accordingly make mistakes, but they seem not to learn anything from them that they can apply in a different situation. If we move the beavers from one enclosure to another, after they have learned not to react to the sound of running water at the inlet, they seem to have to learn the same thing all over again in the new surroundings.

In order to find out whether the ability to react to the different stimuli that activate and control dam-building is inborn, I brought up Findus so that he never saw or heard running water. In his second autumn he lived in a small enclosure in which we placed a large tub full of standing water. One end of this was covered with chicken netting so that he would be able to fix sticks in it, but this he had, of course, no reason to do and did not do. But after a time I placed a loudspeaker behind a hole in the side of the tub above the water level.

When I had got everything ready for the experiment I felt a little embarrassed. It was surely naïve to imagine that I could get Findus to build against a wooden board in still water simply by playing for him the sound of rushing water. But if my theories were correct, it would come off. And Findus actually started building to the loudspeaker, as soon as I switched on the player. After that he built every night I played to him. As I have said before, beavers react to unevennesses in the crown of the dam and on the upstream side of the dam, and in the beginning Findus would build to pieces of black paper if I stuck them close to the source of the sound.

Three weeks later I left a length of hose in the tub overnight, with the water turned on. Until then I had been careful to fill it up with water only in the daytime while Findus was asleep, but on this night the water was splashing into the tub close to the end where Findus had not built before. As I did not have the tape recorder on, Findus built only because of the running water, and his activities had a positive result in that he found a way over the pile of building material directly up to the lodge, which was on that side of the tub. After that he went on building only at the new place, and no longer interested himself in any sort of stimuli from the unnecessary dam at the other end of the tub. So acquired learning plays a certain part in the localization of dam-building, and even the choice of material is determined to some extent by the animal's experiences, if supply is limited. If there is plenty of all kinds of material the animal builds with whatever he happens to come across, and this means that there is an adequate variety of suitable kinds without his needing to learn what is suitable and what is not.

Beavers go on for a long time inspecting those parts of the dam where the building work was first activated. They seem to remember them long after the activating stimulus has been removed. They have also shown in other ways that they have a very highly developed capacity to learn and remember the position of various details inside their territory. Even when they are on the ground, for example, they seem to know where their passages run, deep under the surface.

In the matter of dam-building, too, the movement patterns and their sequence are inborn. From the beginning the behavior is controlled exclusively by inborn reactions to certain stimuli, though individuals can learn afterward not to react to some of them and to react to others.

In the development of the individual, inborn and acquired reactions are interwoven to form a complicated entity. There is still, of course, a great deal of work to be done before we understand all the details of this entity. But basically all beavers react in the same way and learn the same details in the same environment, and there are great possibilities of carrying these studies further. Such studies are justified, in that the behavior of almost all species of animals seems to be organized according to the same basic principles. It would be strange if human behavior had nothing in common with that of the other animals. No doubt ethology will come by degrees to contribute more and more to the understanding of our own peculiar reactions.

The observations recorded here are only a modest contribution to our knowledge of one species of mammal. I will be satisfied if I have succeeded in showing that the beaver is worthy of our interest from many points of view, and that studies of this sort have many functions to fulfill. It is really surprising how little we know about the living creatures around us, and our respect and regard for them is in direct proportion to our knowledge. Our world is being dominated more and more by technology, and we are going to need greater and greater insight into the living world if our existence is not to become unendurable.

Autumn has just passed over into winter. These past few weeks the beavers in both the big dams have been keeping a hole open in the ice, but last night the nine animals were finally shut in under the cover of ice. The pair in the stream aquarium can go on building all through the winter, and in the small enclosure where Findus lives with his "fiancée" we can keep the water open until it gets really cold.

Findus is now beginning to get more and more attached to his female and no longer stands by the netting "weeping" every time I leave him. But when I go down to the enclosure and switch on the brilliant searchlight his shaggy back soon appears above the coal-black water. He comes toward me with an unmistakable look of joy in his eyes. His movements seem rather heavy and his whole figure takes one's thoughts back to the beginnings of time and untouched virgin lands. It is rather strange to have an intimate relationship with such a creature. We talk and groom one another for a time, then he goes back to his own world under the black sheet of water. His devotion is boundless because it is uncomplicated. I shall presumably be the object of it as long as he lives.

Bibliography

Bailey, V. "Beaver Habits and Experiments in Beaver Culture." *Depart. Agr. Tech. Bull.* 21. 1927.

Barnett, S. A. *A Study in Behaviour*. London: 1963.

Bourlière, François. *Seder och skick i djurvärlden*. Stockholm: 1959.

Bradt, G. W. "A Study of Beaver Colonies in Michigan." *Jour. Mamm.* 19. 2. 1938.

Dezschkin, V. V. *The Beaver's Mating Behavior. Publications of the Voronezsch Reserve* VII. Voronezsch: 1957. (In Russian)

Dugmore, A. R. *The Romance of the Beaver*. London: 1914.

Eibl-Eibesfeldt, I. "Das Verhalten der Nagetiere." *Handbuch der Zoologie* 10 (13) 8, 12. 1958.

Fabricius, E. *Etologi*. Stockholm: 1961.

Fries, C. *Bäverland*. Stockholm: 1960.

Hinze, G. *Der Biber*. Berlin: 1950.

Martin, H. T. *Castorologia*. 1892.

Morgan, L. H. *The American Beaver and His Works*. Philadelphia: 1868.

Osborn, D. J. "Age Classes, Reproduction and Sex Ratios of Wyoming Beaver." *Jour. Mamm.* 34. I. 1953.

Scott, John Paul. *Djurens beteende*. Stockholm: 1962.

Shadle, A. R. "The American Beaver." *Animal Kingdom*. Vol. LIX No. 4. 1956.

Wilsson, L. "Inventering av bäverstammen vid Faxälven med biflöden inom Ångermanland." *Sv. J.* 6. 1959.

——— "Fångst och omplantering av bäver." *Sv. J.* 4. 1960.

——— "Om bäverinventering." *Sv. J.* 4. 1961.

——— "Om bävrarna vid Faxälvens vattensystem." *Zool. Revy* 2. 1960.

——— "Medfödda och förvärvade beteendemönster hos bävern." *Svensk Naturvetenskap*. 1962.

——— *Observations on the behavior of the European beaver* (*Castor fiber* L.). (Being revised)

Yeager, L. E. and Rutherford, W. H. "An Ecological Basis for Beaver Management in the Rocky Mountain region." *Transactions of the second N. A. Wildlife Conference*, March 4, 5, and 6, 1957.

Index

Acquired behavior, 124, 148, 149. *See also* specific kinds
Aggressiveness, 62–63, 74–75, 81, 96, 131, 133; learning of, 138
Air (oxygen), 15, 30. *See also* Ventilation holes
American beaver, v, x, 2, 4–5, 6–8
Angermanälven, 83
Angermanland, 7, 24, 100
Angermann River, 70
Aquarium beaver experiments, 109–22
Artificial beaver streams, 109–22
Asia, 2
Aspens, 57, 80, 101; felling, 102, 105–6
Autumn activity, 113, 129, 141, 142. *See also* specific kinds

Baby beavers, 42–53, 54–69, 70–79, 135–38 (*See also* Birth process; Family life; Mating; specific behavior patterns); and learning, 139–50
Bark, 11, 16, 26, 49–50, 52, 56, 95, 143; aspen, 105–6, 140–41
Bathing, 46, 51, 56, 92, 135
Beaver Club, 5
"Beaver economy," 2, 3, 4
Behavior patterns, 123–27. *See also* Learning; specific behavior patterns
Biology Institute, 54, 62, 63
Birth process, 42, 85–98, 129
Birth rate, 134
Biting, 75
Body weight, beaver, v, 15, 46, 88, 141
Bones, beaver, 2
Building activity. *See* Dams; Lodges
Burrows, 80

Canada, 4, 8
Castor canadensis (American beaver), v, x, 2, 4–5, 6–8
Castoreum, v, 4, 61, 97; depositing of, 38, 39–40, 41, 131–32; described, x; medicinal use, x, 2, 3; significance of, 38, 39, 40, 61, 63, 131–32
Castor fiber (European beaver), v, x, 3–4, 7
Civilization, beavers and, 3–4
Claws, 51–52
Collecting of stores, 19–23, 27–31, 53, 76, 83; learning, 141, 143–44
Contact, body, 138
Contact sounds, 62, 87, 92, 94, 95, 97, 134, 138
Cooperative behavior, 133–34, 135, 136. *See also* specific kinds
Courting, 132. *See also* Mating

Dams; dam-building, 1, 13, 76, 77, 79, 145–49; aquarium, 109–22; learning, 145–49; summer, 102; winter, 80–84
Dancing, 44–45, 102
Deciduous trees, 18, 19, 102. *See also* specific kinds
Diarrhea, 126
Diet. *See* Feeding behavior
Digging behavior, 64–65, 71–72, 141, 143
Diving, 95, 113–15; speed, 116
Dwelling holes. *See* Lodges

Eating. *See* Feeding behavior
Elbe, 4, 42
England, 3, 4–5
Environment, adjustment to, 123–27. *See also* Learning
Eriksson, Alfons, 33, 42, 50
Eriksson, Elias, xi, 1, 9–23, 24–33, 42, 109
Erosion, land, 6
Esmeralda (beaver), 66–68
Europe, early: beavers in, 3, 4, 5
European beaver, v, x, 3–4
Eva (beaver), 74, 81–82
Excreta, eating of, 48–49
Extermination of beavers, 4–6, 133
Eyes, expressiveness of, 138

Fabricius, Eric, 42
Fabricius, Pian, 42
Family life, vi, 40, 42, 85–97, 128–38. *See also* Female beavers; Male beavers; specific aspects
Faxälven beavers, xi, 9, 17, 18, 19, 24, 25, 29, 32, 33, 125–26; river lodges, 35, 41
Feeding behavior, v–vi, 11, 12, 15–18, 35–37, 39, 40, 101, 102, 133 (*See also* Collecting of stores); competition and, 129; learning of, 143–44; winter stores, 19–23, 26, 27, 28; young beavers and, 46, 48–50, 56–58, 70–71, 73, 94, 126, 143
Feet, 114–15
Female beavers, 85–98 (*See also* Family life); and dam-building, 118–19
Fighting. *See* Aggressiveness
Fina (beaver), 62, 66–68
Findus (beaver), 123, 126, 137, 140–42, 148, 150
Finland, 7
Fir trees, 23
Fish; fishing, 33, 68–69, 99–100, 108, 109
Flight behavior, 137, 139–40
Flight reactions, 137, 140
Floods, 18
Food. *See* Feeding behavior
Foxes, 100
France, 4, 5
Fur. *See* Skins, beaver

Gamekeepers' Association, xii, 50
Game Preservation Authority, 21
Gestation period, 135
Gnawing, 56–58
Greta (beaver), 74, 81–82, 109, 112, 113, 122
Grooming, 37–38, 45–46, 51–52, 56, 71, 101; baby beavers, 90–91, 96, 126, 140; learning, 140; mutual, 138
Group cooperation, 133–35
Gulls, 99

Härnösand, 24
Hearing sense, 60
Hearne, Samuel, x
Hediger, Professor, 135
Helgum, Lake, 18, 24, 26, 28
Hippocrates, x
History, beaver, ix–xii, 1–6
Hölle laboratory, 74, 78, 81, 109, 122
Hollows, as stimuli for building activity, 113, 143
Holmeselet, 10–13, 16, 19, 20, 28, 29, 32, 33; timber-dumping ground on, 68
Homes. *See* Lodges
Hudson's Bay Company, 4–5
Hunting, ix, 2–3, 4–6, 133

Ice, beaver activity and, 13–15, 18–23, 29, 30, 31, 80–84
Imprinting, 137–38
Inborn behavior, xi, 124, 140–50
Incisors, 57
Indalsälven, 74, 109
Indians, North American, ix, x, 2, 4–5
Inherited tendencies, xi, 124, 140–50
Italy, 3

Jämtland, 7
Jämtland-Angermanland, 7

Land erosion, 6
Learning, 137, 139–50. *See also* Baby beavers; specific aspects, behavior patterns
Leaves, 49–50, 52, 70–71, 141
Life expectancy of beavers, 134
Linnaeus, Carolus, x
Lodges (lodge-building), vi, 11–18, 24ff., 82–83, 132; aquarium, 111; learning, 141–50; river lodges, 34–36, 76; young beavers and, 58–61, 141–50

Male beavers, 99–108 (*See also* Family life; specific behavior activity); and dam-building, 118–19
Man, primitive: beavers and, 2–3
Maternal behavior, 87–98, 128–38. *See also* Pregnancy
Mating, 85–98, 112, 123, 128, 129, 132, 134–35
Medicinal uses of castoreum. *See under* Castoreum
Mediterranean area, 3
Milk, beaver, 126
Minks, 100
Mohlén, Mr. (caretaker), 57, 62, 67
Molars, 56, 57
Montreal, 4, 5
Movement patterns, 123, 138, 140. *See also* specific kinds

Natural selection, 144–45
Norrbotten, 7
Norrland forests, 4, 8, 9
North-West Company of Montreal, 5
Norway, 4, 7
Number Five (beaver), 55, 73–74
Number Four (beaver), 55, 73–74
Nursery period, 128ff., 139, 140, 143

Otters, 100

Pairing, 129–32, 134–35
Passages, construction of, 12–15, 18, 25; in river lodges, 34–36; young beavers and, 72, 143, 144
Placenta, 88, 89, 135
Plastering, 16, 76, 144, 145
Play; playfulness, 95, 102, 136–37, 138
Poland, 4
Po River, 3
Post, Tall von, xii, 24, 26, 28, 30, 32, 42, 44; as Game Preservation adviser, 21
Pregnancy, 86–98, 125–26, 135. *See also* Maternal behavior

Ramsele, xi, 9, 24, 32; beaver lodges in, 17, 18, 25, 26
Regulations, water, 19–23, 32, 33
Reservoirs, 33, 68
Rhône delta, 4
Riparian Court, 127
River lodges, 34–36, 76
River Regulation Board, 32, 33
Roofs, beaver lodge, 35, 36, 37, 143
Russia. *See* Soviet Union

Salmon, 99, 108, 109
Salmon Breeding Institute, 109
Seals, 100, 108
Sex: castoreum depositing and, 38; determination of, 7, 55, 81; family life and, 128
Siberia, 4, 7
Skansen (Stockholm, Sweden), 83
Skins, beaver, 2, 3, 4
Sleeping chambers, 35, 60–61, 113
Sociability, 133
Social behavior and organization. *See* Family life; specific behavior patterns
Sollefteå, 1, 83
Soviet Union, 2, 4, 5, 7, 8; beaver studies by, 34, 42, 133, 135
Spain, 3
Spring activity, 129, 136. *See also* specific kinds of activity
Stina (beaver), 74, 81–82, 109, 112, 113, 122
Stockholm, 83
Stones, 117, 146
Stores, collecting of, 19–23, 27–31, 53, 76, 83; learning process, 141, 143–44
Studies, beaver, 34–41, 42, 133, 135
Suckling, 88, 95
Summer activity, 18, 70–78. *See also* specific kinds
Sweden, 3, 7, 18, 100. *See also* specific places
Swimming, v, 45, 94, 113–16, 136; learning, 136, 141; speed, 115

Tail, v, 115
Tail beat, 113–14
"Talking," 46, 111, 138. *See also* Contact sounds
Teeth, v, 2, 56–57
Terrarium, use of, 54ff.
Territorial behavior, 40, 45, 50, 65, 71, 82, 83; castoreum deposits and, 38, 39–40, 61, 63, 132; learning of, 141
Traps, beaver, 50
Trees (tree-felling), 11, 12–16, 18, 19, 26, 78, 83, 101–7. *See also* specific kinds
Tryggve (Stockholm schoolboy), 70, 71, 72, 73, 74
Tuff (beaver), 55ff., 59–63, 67, 71–79; and dam-building, 77; death of, 125; and digging, 65–66; and maternity, 85–96, 124–25
Turnips, 80, 97
Tuss (beaver), 55ff., 71–74, 77, 78–79, 125–26; and mating, 85–96, 125

Vagnforsen, 32, 126
Värmland-Dalecarlia, 7
Ventilation holes, 15, 30, 36; winter, 80

Wanderlust, 129

Water flow, dam-building activity as reaction to, 117–22, 145–48
Water level, 68–69, 75, 76; dam-building and, 118–22; lodge construction and, 36, 83; short-term regulations, 18–33, 127
Water supply, role of beaver in maintaining, 6
Weight. *See* Body weight
Winter activity, 19–33, 80–84, 126–27. *See also* specific kinds
Winter stores, 19–23, 27–31; young beavers and, 53, 57, 77–78
Wrestling, 45, 62, 131

X-rays, use of, 55, 74, 86–87